T0133643

Hybrid Computational Intelligence

Research and Applications

Hybrid Computational Intelligence

Research and Applications

Siddhartha Bhattacharyya
Václav Snášel
Indrajit Pan
Debashis De

CRC Press
Taylor & Francis Group
Boca Raton London New York

CRC Press is an imprint of the
Taylor & Francis Group, an **informa** business

CRC Press
Taylor & Francis Group
6000 Broken Sound Parkway NW, Suite 300
Boca Raton, FL 33487-2742

Printed on acid-free paper

International Standard Book Number-13: 978-1-1383-2025-3 (Hardback)

Visit the Taylor & Francis Web site at
http://www.taylorandfrancis.com

and the CRC Press Web site at
http://www.crcpress.com

Siddhartha Bhattacharyya would like to dedicate this book to his father, the late Ajit Kumar Bhattacharyya, his mother, the late Hashi Bhattacharyya, his beloved wife, Rashni, and his revered research supervisor, Prof. (Dr.) Ujjwal Maulik.

Václav Snášel would like to dedicate this book to his wife, Božena Snášelová.

Indrajit Pan would like to dedicate this book in memory of his grandfathers, the late Anath Bandhu Ghosh, and the late Amal Bandhu Ghosh, for their unconditional love, support and guidance throughout.

Debashis De would like to dedicate this book to his grandfather, Sri Sadananda De, and grandmother, Smt. Belu Rani De.

Contents

Preface **xiii**

Authors **xvii**

1 Nature-Inspired Algorithms: A Comprehensive Review **1**

Essam H. Houssein, Mina Younan, and Aboul Ella Hassanien

1.1 Introduction 2
1.2 Research Trends 2
1.2.1 Based on Algorithm Idea 3
1.2.2 Based on Problem Type 3
1.2.3 Based on Algorithm Applications 3
1.3 Classification of Nature-Inspired Algorithms 4
1.3.1 SI-Based Algorithms 5
1.3.2 BI-not-SI-Based Algorithms 5
1.3.3 Natural Science-Based Algorithms 6
1.3.4 Natural Phenomena-Based Algorithms 7
1.4 Variants of Nature-Inspired Algorithms 7
1.4.1 Binary Algorithms 7
1.4.2 Chaotic Algorithms 7
1.4.3 Multi-objective Algorithms 8
1.4.4 Hybrid Algorithms 9
1.5 A Review of the Most Recent NI Algorithms 9
1.5.1 Artificial Butterfly Optimization Algorithm 9
1.5.2 Grasshopper Optimization Algorithm 10
1.5.3 Salp Swarm Optimization Algorithm 12
1.5.4 Spotted Hyena Optimization Algorithm 14
1.5.5 Chemotherapy Science Optimization Algorithm 15
1.6 Conclusion 17

2 Hybrid Cartesian Genetic Programming Algorithms: A Review **27**

Johnathan Melo Neto, Heder S. Bernardino, and Helio J.C. Barbosa

2.1 Introduction 28
2.2 Metaheuristics 30
2.2.1 Single-Solution Methods 31

2.2.2 Population-Based Methods 31
2.2.2.1 Evolution strategies 31
2.2.2.2 Differential evolution 32
2.2.2.3 Biogeography-based optimization 32
2.2.2.4 Non-dominated sorting genetic algorithm . . 33
2.2.2.5 Harmony search 34
2.2.2.6 Estimation of distribution algorithms 35
2.2.2.7 Ant colony optimization 35
2.2.2.8 Particle swarm optimization 36
2.3 Fundamentals of Cartesian Genetic Programming 37
2.3.1 Historical Context 37
2.3.2 Encoding 37
2.3.3 Evolution Scheme 38
2.3.4 Parameters 39
2.3.5 Advantages and Drawbacks 39
2.4 Literature Review on Hybrid Metaheuristics 40
2.5 Hybrid Cartesian Genetic Programming Algorithms 42
2.5.1 Motivation 42
2.5.1.1 CGP combined with ant colony optimization 42
2.5.1.2 CGP combined with biogeography-based optimization and opposition-based learning . 44
2.5.1.3 CGP combined with differential evolution . . 45
2.5.1.4 CGP combined with estimation of distribution algorithm 47
2.5.1.5 CGP combined with NSGA-II 48
2.5.1.6 CGP combined with harmony search 49
2.5.1.7 CGP combined with particle swarm optimization 51
2.6 Discussion on Hybrid CGP Algorithms 52
2.7 Future Directions of Hybrid CGP Algorithms 52
2.8 Concluding Remarks 55

3 Tuberculosis Detection from Conventional Sputum Smear Microscopic Images Using Machine Learning Techniques 63
Rani Oomman Panicker, Biju Soman, and M.K. Sabu
3.1 Introduction 63
3.2 Sputum Smear Microscopic Images 65
3.2.1 Disadvantages of Conventional Methods 66
3.3 Machine Learning Techniques for TB Detection 67
3.3.1 Study Design 68
3.3.2 Literature Review 68
3.4 Discussion 76
3.5 Conclusions and Future Scope 77

4 Privacy towards GIS Based Intelligent Tourism Recommender System in Big Data Analytics **81**

Abhaya Kumar Sahoo, Chittaranjan Pradhan, and Siddhartha Bhattacharyya

4.1 Introduction 82
4.2 Background 83
4.2.1 Intelligent Tourism Recommender System and its Basic Concepts 84
4.2.2 Phases of Tourism Recommender System 84
4.2.3 Collaborative Filtering Technique Used in TRS 85
4.2.3.1 Memory-based collaborative filtering 86
4.2.3.2 Model-based collaborative filtering 87
4.2.3.3 Evaluation of TRS 88
4.3 Geographical Information System Used in TRS 90
4.4 Big Data Analytics in Tourism 90
4.5 Machine Learning Techniques Used in GIS-based TRS 92
4.6 Privacy Preserving Methods Used in GIS-based TRS 94
4.7 Proposed Privacy Preserving TRS Method Using Collaborative Filtering 97
4.7.1 Dataset Description 97
4.7.2 Experimental Result Analysis 97
4.8 Conclusion and Future Work 97

5 Application of Artificial Neural Network: A Case Study of Biomedical Alloy **101**

Amit Aherwar and Amar Patnaik

5.1 Introduction 102
5.2 Test Material and Methods 105
5.2.1 Test Materials 105
5.2.2 Manufacturing of Orthopaedic Material 105
5.2.3 Material Characterization 107
5.2.4 Mechanical Studies 107
5.2.5 Wear Measurement of Orthopaedic Material 107
5.2.6 Taguchi Design of the Experiment 109
5.3 Results and Discussions 110
5.3.1 Phase Analysis and Microstructure 110
5.3.2 Mechanical Studies of Manufactured Material 111
5.3.2.1 Micro-hardness 111
5.3.2.2 Compressive strength 113
5.3.3 Taguchi Experimental Design 113
5.4 Simulation Model for Wear Response 114
5.4.1 Data Processing in ANN Model 116
5.4.2 Network Training 117
5.4.3 Neural Network Architecture 120
5.4.4 ANN Prediction and its Factor 120
5.5 Conclusion 124

6 Laws Energy Measure Based on Local Patterns for Texture Classification 131

Sonali Dash and Manas R. Senapati

6.1 Introduction . . . 131
6.2 Related Work . . . 134
6.2.1 Mathematical Background of LBP . . . 134
6.2.2 LBP Minimum . . . 136
6.2.3 LBP Intensity . . . 136
6.2.4 LBP Uniform . . . 136
6.2.5 LBP Number . . . 137
6.2.6 LBP Median . . . 137
6.2.7 LBP Variance . . . 138
6.2.8 CLBP . . . 138
6.2.9 Sobel-LBP . . . 139
6.2.10 Laws' Mask . . . 140
6.3 Local Pattern Laws' Energy Measure . . . 140
6.3.1 Problem Formulation . . . 140
6.4 Implementation and Experiments . . . 142
6.4.1 Results of Brodatz Database . . . 145
6.4.2 Results of ALOT Database . . . 146
6.4.3 Statistical Comparison of the Methods . . . 148
6.5 Conclusion . . . 149

7 Analysis of BSE Sensex Using Statistical and Computational Tools 153

Soumya Chatterjee and Indranil Mukherjee

7.1 Introduction . . . 154
7.2 The Data Analysed . . . 156
7.2.1 Return and Raw Data . . . 156
7.2.1.1 Time series of the return data created from raw data . . . 157
7.2.1.2 Return data created from detrended data . . . 158
7.2.1.3 The role of raw data in analyses . . . 158
7.3 The Data Vectors and Principal Component Analysis . . . 158
7.3.1 Construction of the Data Vectors . . . 159
7.3.2 Principal Component Analysis . . . 160
7.3.3 PCA of Raw Sensex Data . . . 160
7.3.4 PCA of Detrended Sensex Data . . . 161
7.3.5 PCA of Raw Sensex Data with Noise . . . 161
7.3.6 PCA of Return Sensex Data . . . 162
7.3.6.1 PCA of the return data obtained from raw Sensex data . . . 162
7.3.6.2 PCA of the return data from detrended Sensex data . . . 163

7.4 Kernel Principal Component Analysis 165
7.4.1 KPCA of Raw Sensex Data 165
7.4.1.1 Methodology 166
7.4.1.2 Results of KPCA applied to raw Sensex data (with trend) 166
7.4.2 KPCA of Raw Trend-Removed Sensex Values 166
7.4.3 KPCA of Raw Sensex Data with Noise 168
7.4.4 KPCA of Return Sensex Data 168
7.4.4.1 KPCA of return Sensex data 168
7.4.4.2 KPCA of return of detrended Sensex data . 170
7.5 Detrended Fluctuation Analysis 172
7.5.1 Detrended Fluctuation Analysis of the Detrended Sensex Data 172
7.6 Conclusion 174

8 Automatic Sheep Age Estimation Based on Active Contours without Edges 177

Aya Abdelhady, Aboul Ella Hassanien, and Aly Fahmy

8.1 Introduction 177
8.2 Related Work 178
8.3 Theory and Background 180
8.3.1 Active Contours 180
8.3.2 Blob Detection and Counting 181
8.3.3 Morphological Operations 181
8.3.4 Dentition 181
8.3.5 Image Collection and Camera Setting 181
8.4 The Proposed Automatic Sheep Age Estimation System 182
8.4.1 Pre-processing Phase 183
8.4.2 Segmentation Phase 183
8.4.3 Post-processing Phase 185
8.4.4 Age Estimation Phase 185
8.5 Experimental Results and Discussion 188
8.6 Conclusion and Future Work 192

9 Diversity Matrix Based Performance Improvement for Ensemble Learning Approach 195

Rajdeep Chatterjee, Siddhartha Chatterjee, Ankita Datta, and Debarshi Kumar Sanyal

9.1 Introduction 196
9.2 Related Work 196
9.3 Theoretical Background 197
9.3.1 Wavelet Based Energy and Entropy 197
9.3.2 Ensemble Classification 199
9.3.2.1 Bagging ensemble learning 199

9.3.2.2 Majority voting 200
9.3.3 Used Diversity Techniques 200
9.3.3.1 Cosine dissimilarity 201
9.3.3.2 Gaussian dissimilarity 202
9.3.3.3 Kullback-Leibler divergence 202
9.3.3.4 Euclidean distance 202
9.4 Proposed Method 202
9.5 Results and Discussion 205
9.5.1 Preparing the Used Datasets 205
9.5.2 Experimental Set-up 207
9.5.3 Results Analysis 208
9.6 Conclusion and Future Work 211

Index **217**

Preface

Imparting intelligence to machines has been one of the challenging thoroughfares in the computer vision research community. A plethora of classical approaches have been invented to solve real-world problems in science, engineering, business and medicine. But most of these approaches suffer from imprecision and inaccuracy in some respects. In addition, almost each and every event in this workaday world exhibits a huge amount of uncertainty and ambiguity, which the classical approaches fail to address to any extent. Hence, scientists have come up with some intelligent techniques which can come to the rescue. These techniques have been found to perform better in handling uncertainties in existing problems. However, when these techniques are conjoined together, they culminate into more efficient and robust systems, even in highly unpredictable circumstances. These hybrid computational intelligent techniques have been found very promising in dealing with the real world problems encountered in different fields of engineering. The core message is that this book is a synergistic integration of hybrid computational intelligent techniques applied to engineering problems.

This book aims to provide an exhaustive review of the hybrid computational intelligent paradigm, with supportive case studies. In addition, it also aims to provide a gallery of engineering applications where this computing paradigm can be effectively put to use.

This volume comprises nine well-versed chapters reporting overview and latest trends in applications of hybrid computational intelligence.

Natural inspired algorithms mimic different features from the nature for optimizing results and solutions. Features like diversity and dynamicity of the natural phenomena are great sources for inspiring these algorithms in order to provide such intelligent systems. The natural inspired algorithms also use a balance between exploitation and exploration for this purpose. Chapter 1 presents a review of the most popular nature-inspired optimization algorithms in vogue with reference to the future trends in evolving advanced incarnations of these algorithms.

Genetic Programming (GP) is a type of evolutionary algorithm that has been applied to many real-world problems in computer science, engineering, and other fields. Particularly, Cartesian Genetic Programming (CGP) is a GP approach which uses a graph representation to encode programs. Chapter 2 proposes a critical review of CGP algorithms that make use of hybrid approaches. A detailed description of CGP and suggestions on how it can be combined with other search mechanisms is presented.

Tuberculosis (TB) is an infectious airborne disease particularly affecting the lungs. It is caused by a rod-shaped bacterium called Mycobacterium tuberculosis and is a leading cause of death. Most of the developing countries follow sputum smear microscopic examination for TB detection, but it is labor-intensive and error-prone. In Chapter 3, the authors cover a wide spectrum of machine learning techniques that are used for detecting TB bacilli from conventional sputum smear microscopic images. The authors analysed all the studies published between 2008 and 2018 by using a search strategy and present here TB detection using machine learning techniques only 16 relevant studies based on machine learning techniques.

Machine learning algorithms are often used in designing of intelligent based Tourist Recommender Systems (TRS) to improve decision making, clustering, knowledge representation and planning. In Chapter 4, the authors compare three different privacy preserving methods such as Nave Bayesian classification, non-negative matrix factorization and randomization method, etc. used in collaborative filtering based TRS to provide security to sensitive information related to user and tourist locations. They propose a Singular Value Decomposition (SVD) based privacy preserving method with Pearson's coefficient distance to achieve better accuracy while preserving privacy in TRS as compared to the three methods discussed.

Cobalt-Chromium based alloys are widespread materials used in medical industry. Despite their success in many applications, some problems such as wear still persist. To avoid these, the constituents of Co-Cr-Mo alloy should be formulated with accurate proportions. In Chapter 5, a series of biomedical alloys containing different amounts of nickel were manufactured by induction furnace. Moreover, an artificial neural network model assisted by Levenberg-Marquart (LM) learning algorithm has been presented for the prediction of the effects of the nickel on tribology properties of fabricated Co-30Cr-4Mo alloys for orthopaedic material. The results of the developed model shows a healthy conformity with the experimental results.

Chapter 6 uses the Laws' mask method for texture analysis. Laws' masks are used to detect the edge, level, spot, ripple and wave within an image, but they do not take care of the variation in illumination and robustness of noise in an image. Based on these features, the authors propose a method called Local Pattern Laws' Energy Measure (LPLEM) that incorporates Local Binary Pattern (LBP) and LBP variants with Laws' Mask for feature extraction. Eleven feature extraction techniques are employed by combining different LBP variants with Laws' mask technique in the experiment. To prove the effectiveness of the proposed method, experiments have been conducted on two challenging (Brodatz and ALOT) databases. The best model is derived from a classification process effected by the k-NN classifier. The proposed LPLEM method is compared with the existing Laws' mask descriptor. Experimental results demonstrate the efficiency of the suggested approach.

Chapter 7 analyses the different features of the Indian financial market using intraday prices covering the years 2006-2012, treating financial index as signal from the stock market considered as a complex system. Different tools from statistical physics and computational sciences are used in the analysis. These tools include Principal Component Analysis (PCA), Kernel Principal Component Analysis (KPCA), Hurst exponents, Detrended Fluctuation Analysis (DFA), etc. Both raw and return data are analysed, including their original and detrended values. A high degree of correlation is observed in the values, this correlation being independent of the data size or the year considered for analysis. The pattern of movement of the Sensex data is observed to be nearly unidirectional.

A knowledge about the age of sheep is necessary in the evaluation of growth performance and in deciding which individuals can be chosen to buy, sell, cull and mate. However, it is hard to keep age records in farms. Therefore, farmers depend on dentition to estimate sheep age, but most people do not have that required knowledge for age estimation. Chapter 8 introduces a fully automated technique for real-time sheep age estimation. Sheep age has been estimated through dentition automation by recognizing the size and number of sheep teeth. In the proposed approach, teeth were segmented using active contours without edges for multiple objects extraction. Then the detected connected components of the segmented teeth were counted and their dimensions were measured. The proposed approach can also be used to estimate the age for other ruminant animals. In this work, sheep age estimation techniques were tested on a data set of 52 teeth images with sufficient good accuracy.

The ensemble learning approach is one of the widely used machine learning technique in classification problems. There are different types of ensemble classifiers and their combining techniques in use. In Chapter 9, a bagging ensemble technique is used on a motor-imagery electroencephalography (EEG) signal for brain-states discrimination problem. The majority-voting technique is implemented at the end for combining different predictors and to compute the final decision. It is observed that diversity in decision boundaries obtained from different learners plays an important role in the performance of the ensemble. The authors have also proposed a new diversity matrix based pruning technique which selects a subset of predictors from the actual set of predictors obtained from the ensemble. It has been found that the proposed method performs quite satisfactory compared to existing methods.

This volume is intended for the students of computer science, electrical engineering, information sciences of different universities to cater to a part of their curriculum. The editors would find this venture fruitful if this volume comes to use for the academic and scientific community as well. The editors would like to take this opportunity to render their heartfelt gratitude to all the contributory authors and reviewers for their cooperation during the book project. Thanks are also due to Dr. Gagandeep Singh, Editorial Manager and

Senior Editor (Engineering/Environmental Sciences) at CRC Press for his support and cooperation during the book project.

February 2019
India and the Czech Republic

Siddhartha Bhattacharyya
Václav Snášel
Indrajit Pan
Debashis De

MATLAB® is a registered trademark of The MathWorks, Inc. For product information please contact:

The MathWorks, Inc.
3 Apple Hill Drive
Natick, MA, 01760-2098 USA
Tel: 508-647-7000
Fax: 508-647-7001
E-mail: info@mathworks.com
Web: www.mathworks.com

Authors

Dr. Siddhartha Bhattacharyya did his Bachelors in Physics, Bachelors in Optics and Optoelectronics and Masters in Optics and Optoelectronics from University of Calcutta, India in 1995, 1998 and 2000 respectively. He completed PhD in Computer Science and Engineering from Jadavpur University, India in 2008. He is the recipient of the University Gold Medal from the University of Calcutta for his Masters. He is the recipient of the coveted National Award Adarsh Vidya Saraswati Rashtriya Puraskar for excellence in education and research in 2016. He is the recipient of the Distinguished HoD Award and Distinguished Professor Award conferred by Computer Society of India, Mumbai Chapter, India in 2017. He received the Honorary Doctorate Award (D. Litt.) from The University of South America and the South East Asian Regional Computing Confederation (SEARCC) International Digital Award ICT Educator of the Year in 2017. He also received the Rashtriya Shiksha Gaurav Puraskar from Center for Education Growth and Research, India in 2017. He has been appointed as the ACM Distinguished Speaker for the tenure 2018–2020. He received the Young Scientist (Science & Technology) Award from CSERD, India in 2018.

He is currently serving as the Principal of RCC Institute of Information Technology, Kolkata, India. In addition, he is also serving as the Professor of Computer Application and Dean (Research and Development) of the institute. He served as a Senior Research Scientist in the Faculty of Electrical Engineering and Computer Science of VSB Technical University of Ostrava, Czech republic. Prior to this, he was the Professor of Information Technology of RCC Institute of Information Technology, Kolkata, India. He served as the Head of the Department from March, 2014 to December, 2016. Prior to this, he was an Associate Professor of Information Technology of RCC Institute of Information Technology, Kolkata, India from 2011–2014. Before that, he served as an Assistant Professor in Computer Science and Information Technology of University Institute of Technology, The University of Burdwan, India from 2005–2011. He was a Lecturer in Information Technology of Kalyani Government Engineering College, India during 2001–2005. He is a co-author of 5 books and the co-editor of 30 books and has more than 220 research publications in international journals and conference proceedings to his credit. He has got two PCTs to his credit. He was the convener of the AICTE-IEEE National Conference on Computing and Communication Systems (CoCoSys-09) in 2009. He was the member of the Young Researchers'

Committee of the WSC 2008 Online World Conference on Soft Computing in Industrial Applications. He has been the member of the organizing and technical program committees of several national and international conferences. He served as the Editor-In-Chief of International Journal of Ambient Computing and Intelligence (IJACI) published by IGI Global, Hershey, PA, USA from 17th July 2014 to 06th November 2014. He was the General Chair of several international conferences like ICCICN 2014, ICRCICN 2015, ICRCICN 2016, ICRCICN 2017, ICACCP 2017, ICRCICN 2018, ISSIP 2018, ICICC 2018, ICRCICN 2019, ISSIP 2019 and ICACCP 2019.

He is the Associate Editor of International Journal of Pattern Recognition Research and the founding Editor in Chief of International Journal of Hybrid Intelligence; Publisher: Inderscience. He is the member of the editorial board of International Journal of Engineering, Science and Technology and ACCENTS Transactions on Information Security (ATIS). He is also the member of the editorial advisory board of HETC Journal of Computer Engineering and Applications. He is the Associate Editor of the International Journal of BioInfo Soft Computing, IEEE Access and Evolutionary Intelligence, Springer. He is the member of the editorial board of Applied Soft Computing, Elsevier, B. V. He was the Lead Guest Editor of the Special Issue on Hybrid Intelligent Techniques for Image Analysis and Understanding of Applied Soft Computing, Elsevier, B. V. He was the Lead Guest Editor of the Special Issue on Computational Intelligence and Communications in International Journal of Computers and Applications (IJCA); Publisher: Taylor & Francis, UK in 2016. He is the Editor of International Journal of Pattern Recognition Research since January 2016. He was the General Chair of the 2016 International Conference on Wireless Communications, Network Security and Signal Processing (WCNSSP2016) held during June 26–27, 2016 at Chiang Mai, Thailand. He was also the General Chair of Second International Conference on Innovative Computing and Communications (ICICC 2019) held during March 21–22, 2019 at Ostrava, Czech Republic. He is serving as the Series Editors of IGI Global Book Series Advances in Information Quality and Management (AIQM), De Gruyter Book Series Frontiers in Computational Intelligence (FCI), CRC Press Book Series Computational Intelligence and Applications, Wiley Book Series Intelligent Signal and Data Processing and Elsevier Book Series Hybrid Computational Intelligence for Pattern Analysis and Understanding.

His research interests include soft computing, pattern recognition, multimedia data processing, hybrid intelligence and quantum computing. Dr. Bhattacharyya is a life fellow of Optical Society of India (OSI), India, life fellow of International Society of Research and Development (ISRD), UK, a fellow of Institute of Electronics and Telecommunication Engineers (IETE), India and a fellow of Institution of Engineers (IEI), India. He is also a senior member of Institute of Electrical and Electronics Engineers (IEEE), USA, International Institute of Engineering and Technology (IETI), Hong Kong and Association for Computing Machinery (ACM), USA. He is a life member of Cryptology

Research Society of India (CRSI), Computer Society of India (CSI), Indian Society for Technical Education (ISTE), Indian Unit for Pattern Recognition and Artificial Intelligence (IUPRAI), Center for Education Growth and Research (CEGR), Integrated Chambers of Commerce and Industry (ICCI), and Association of Leaders and Industries (ALI). He is a member of Institution of Engineering and Technology (IET), UK, International Rough Set Society, International Association for Engineers (IAENG), Hong Kong, Computer Science Teachers Association (CSTA), USA, International Association of Academicians, Scholars, Scientists and Engineers (IAASSE), USA, Institute of Doctors Engineers and Scientists (IDES), India, The International Society of Service Innovation Professionals (ISSIP) and The Society of Digital Information and Wireless Communications (SDIWC). He is also a certified Chartered Engineer of Institution of Engineers (IEI), India. He has been inducted in the Board of Directors of International Institute of Engineering and Technology (IETI), Hong Kong.

Václav Snášel is currently the rector of VSB Technical University of Ostrava, Czech Republic. He was responsible investigator and cooperating investigator of 15 research projects in the field of basic and applied research with total amount 65 mil Kc. His research and development experience include over 30 years in the Industry and Academia. He works in a multi-disciplinary environment involving artificial intelligence, bioinformatics, information retrieval, knowledge management, data compression, machine intelligence, neural network, nature and biologically inspired computing, data mining, and applied to various real world problems. Studied numerical mathematics at Palacky University in Olomouc, Ph.D. degree obtained at Masaryk University in Brno, he teaches as a professor at VSB – Technical University of Ostrava. From 2001 to 2009 he worked as a researcher at The Institute of Computer Science of Academy of Sciences of the Czech Republic. Since 2009 he works as head of research programme Knowledge management at IT4Innovation National Supercomputing Center, from 2010 he works as dean of the Faculty of Electrical Engineering and Computer Science. He has given 16 plenary lectures and conference tutorials in these areas. He has authored/co-authored several refereed journal/conference papers and book chapters. He has published more than 500 papers (340 papers are indexed at Web of Science, 560 indexed at Web of Science at Scopus). He has supervised many Ph.D. students from the Czech Republic, Jordan, Yemen, Slovakia, Ukraine, Russia, India, China, Lybia and Vietnam. He also supervised postdoc students from the Slovak Republic, Uruguay and Egypt.

Indrajit Pan is presently an Associate Professor in the Department of Information Technology at RCCIIT, Kolkata. He did his Bachelors in Computer Science and Engineering, Masters in Information Technology and Ph. D. (Engineering) in Information Technology. He was the recipient of *University Medal* in Bengal Engineering and Science University, Shibpur for his for securing *first place* in M. Tech. (IT). Indrajit is the *Member* of *ACM (USA)*, *CSI (India)* and *Senior member* of *IEEE (USA)*. He has edited **Six** books with publishers like *Taylor and Francis*, *Willey*, *Springer Nature*, *De Gruyter*, IGI and **Four** International Conference proceedings with IEEE (USA). He authored about **Forty** research articles so far in different Journals, Book chapters and Conference proceedings. His present research interest includes *Community detection and analysis*, *Information and data diffusion*, *Cloud Computing* and *Microfluidic Biochip*.

Prof. Debashis De earned his M.Tech from the University of Calcutta in 2002 and his Ph.D (Engineering) from Jadavpur University in 2005. He is the Professor and Director in the Department of Computer Science and Engineering of the West Bengal University of Technology, India, and Adjunct research fellow at the University of Western Australia, Australia.He is a senior member of the IEEE. Life Member of CSI and member of the International Union of Radio science. He worked as R&D engineer for Telektronics and programmer at Cognizant Technology Solutions. He was awarded the prestigious Boyscast Fellowship by the Department of Science and Technology, Government of India, to work at the Herriot-Watt University, Scotland, UK. He received the Endeavour Fellowship Award during 2008–2009 by DEST Australia to work at the University of Western Australia. He received the Young Scientist award both in 2005 at New Delhi and in 2011 at Istanbul, Turkey, from the International Union of Radio Science, Head Quarter, Belgium. His research interests include mobile cloud computing, Green mobile networks, and nanodevice designing for mobile applications. He has published in more than 200 peer-reviewed international journals in IEEE, IET, Elsevier, Springer, World Scientific, Wiley, IETE, Taylor Francis and ASP, seventy International conference papers, four researches monographs in springer, CRC, NOVA and ten text books published by Person education.He is Associate Editor of journal IEEE ACCESS, Editor Hybrid computational intelligence.

Chapter 1

Nature-Inspired Algorithms: A Comprehensive Review

Essam H. Houssein
Faculty of Computers and Information, Minia University, Egypt
E-mail:essam.halim@mu.edu.eg

Mina Younan
Faculty of Computers and Information, Minia University, Egypt
E-mail: mina.younan@mu.edu.eg

Aboul Ella Hassanien
Faculty of Computers and Information, Cairo University, Egypt
E-mail: aboitcairo@gmail.com

1.1 Introduction 2
1.2 Research Trends 2
    1.2.1 Based on Algorithm Idea 3
    1.2.2 Based on Problem Type 3
    1.2.3 Based on Algorithm Applications 3
1.3 Classification of Nature-Inspired Algorithms 4
    1.3.1 SI-Based Algorithms 5
    1.3.2 BI-not-SI-Based Algorithms 5
    1.3.3 Natural Science-Based Algorithms 6
    1.3.4 Natural Phenomena-Based Algorithms 7
1.4 Variants of Nature-Inspired Algorithms 7
    1.4.1 Binary Algorithms 7
    1.4.2 Chaotic Algorithms 7
    1.4.3 Multi-objective Algorithms 8
    1.4.4 Hybrid Algorithms 9
1.5 A Review of the Most Recent NI Algorithms 9
    1.5.1 Artificial Butterfly Optimization Algorithm 9
    1.5.2 Grasshopper Optimization Algorithm 10
    1.5.3 Salp Swarm Optimization Algorithm 12
    1.5.4 Spotted Hyena Optimization Algorithm 14
    1.5.5 Chemotherapy Science Optimization Algorithm 15
1.6 Conclusion 17
    Bibliography 18

1.1 Introduction

Solving hard and complex problems is a hot topic in computer science. Dynamic programming solves a small scale of the hard problems presenting exact solutions for them, while metaheuristic algorithms solve a large scale of the highly non-linear optimization problems presenting approximation results for them [1]. Because many of the real-world problems and the NP-hard problems have no optimal solutions, many of the optimization algorithms are used as solution tools for solving a subset of these problems under a set of constraints [2, 3]; No-Free-Lunch theory assures this fact [4, 5]. Metaheuristic algorithms become popular due to their [5] (a) flexibility, where they are considered as a black box, (b) gradient-free mechanism, where they evaluate solutions using cost or objective functions, (c) local optima avoidance, where they are of type stochastic algorithms, i.e., they depend on randomized operators.

This research targets researchers and students who are interested in the field of optimization. This review covers swarm intelligence aspects and applications. The remainder of this research is organized as follows. Section 1.2 briefly focuses the light on current research trends. In Section 1.3, all algorithms are divided into different categories supported with detailed references. It discusses classifications of the nature-inspired algorithms and presents a brief list of algorithms followed by four different variants nature-inspired algorithms in Section 1.4. In Section 1.5, brief details are presented on the latest natural inspired algorithms. Finally, conclusion and future work are discussed in Section 1.6.

1.2 Research Trends

Swarm Intelligence (SI) is a subset of nature-inspired algorithms which focuses on the evolutionary computation [2]. SI provides population based and systematic random search algorithms. SI algorithms are population based and iterative, where population of candidate solutions cooperate among themselves [6] by exchanging information about their fitness and statistically converge toward the global or the optimal solutions over the course of generations [7]. In brief, swarm based algorithms have two different learning capabilities: they balance between discovering the search space with a global search (i.e., exploration) for getting initial candidate solutions and performing the local search (i.e., exploitation) for getting better solutions over the course of generations. Because SI has become a very promising area for research, international conferences are held annually in addition to the journals dedicated to SI [7]. As a result, the number of research works in this field increases continuously. SI research could be categorized into major three parts as follows [7].

1.2.1 Based on Algorithm Idea

Researchers tend to improve execution and enhance results of the optimization algorithms with respect to some criteria such as accuracy, efficiency, convergence speed, problems scale, etc. (e.g., particle swarm optimization (PSO) [8]. Consequences results of studying effectiveness of acceleration coefficients and inertia weight [9] on PSO performance produces all kinds of PSO variants, which have been proposed as new inspired algorithms based on the PSO such as new multi-objective PSO based on decomposition (MPSO/D) [10] and Binary Accelerated PSO (BAPSO) [11].

Hybrid algorithms are new trend in SI [7] which take the advantages and strengths of the combined algorithms (i.e., other SI algorithms) for enhancing their performance in problem solving developing a new algorithm with the new resulting features. There are many ways for integrating swarm algorithms. For instance, in the exploration phase, the PSO algorithm could be used as a global search to discover candidate solutions in the search space, and could also be used as a local search in the exploitation phase for searching for better solutions, e.g., for each sunspot butterfly in a butterfly-inspired algorithm [4] (discussed below in more details), after evaluating the fitness function, apply greedy selection of the original and new locations.

1.2.2 Based on Problem Type

Redeveloping and enhancing swarm algorithms aim to solve and cover new problem types such as unconstrained single-objective problems [7], which are then modified and redesigned to fit and cover other types such as a combination of constrained, unconstrained, single-objective, multi-objective, and combinatorial problems. As a result, "dimensionality curse" of the SI is still a hot research area, where researchers tend to present scalable optimization methods. This field is called "feature selection", which benefits from reducing data dimensionality [12].

1.2.3 Based on Algorithm Applications

Research in SI focuses on addressing and handling various types of problems in the real world [7]. This is due to their characteristics such as searching in unknown search spaces, which may be critical for mathematical optimization methods [9].

In this research, there are about fifty algorithms. Which makes their classification is a tedious and difficult task [6], because there is no standard criterion for classification. Instead, optimization algorithms are classified according to a high level set of selected criteria like main inspiration sources, e.g., biology, physics or chemistry. Thus, deep classification is a very challenging task to be handled in a research paper. In this research work, classification will be focused on the main characteristics (i.e., the objective function) of these algorithms during the development process.

1.3 Classification of Nature-Inspired Algorithms

To a certain extent, new optimization algorithms are developed and modified based on the nature as a main source of inspiration [13]. Therefore, nature-inspired (NI) algorithms have become more popular. Biological characteristics are the dominant characteristics of a large portion of the NI algorithms. Therefore, these are called bio-inspired (BI) (i.e., biology-inspired) [2]. These algorithms are also based on SI. Therefore, they are called SI-based algorithms. Figure 1.1 summarizes the relation between SI, BI, and nature-inspired algorithms.

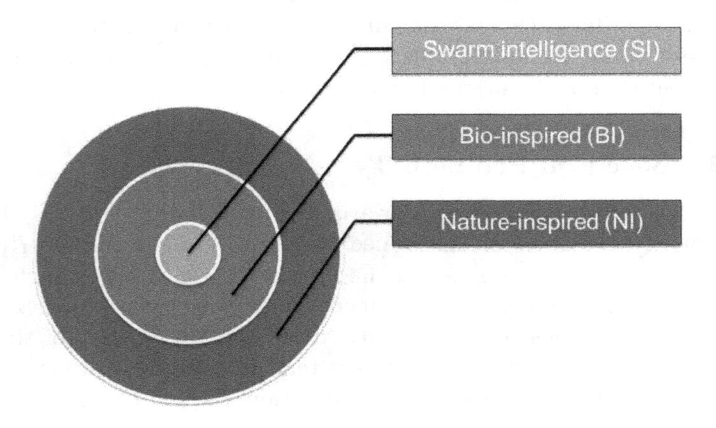

FIGURE 1.1: Relation between NI, BI, and SI algorithms

As mentioned before, there is no unique criterion for classifying optimization algorithms, which makes it a challenging task [6]. Though the most popular classification criteria emphasize the sources of inspiration, but for simplicity, classification could focus the light on the following three points according to its main focus [2]:

1. If the focus is on the search path, then it indicates if algorithms are trajectory-based (i.e., individual based) like simulated annealing (SA) [14] or population based like PSO [8].

2. If the focus is on the interaction of the multiple agents, then it indicates if algorithms are attraction-based like Firefly algorithms (FA) [15] or non-attraction-based like Genetic algorithms (GA) [16].

3. If the focus is on the updating equations, then it indicates if algorithms are equation-based (i.e., use explicit updating equations) like PSO and cuckoo search [17] or they are rule based like GA, which have no explicit updating equations for crossover and mutation operators. There are some algorithms that are rule based and could be converted into equation based like the FA.

Thus the classification process is guided by the actual stimulation and motivation. As we mentioned earlier, optimization algorithms could be categorized into: SI-based algorithms, BI-based algorithms, nature science-based algorithms (NS), and natural phenomena-based algorithms (NPA). All of them will be summarized briefly in the rest of this chapter focusing on relatively new optimization algorithms.

1.3.1 SI-Based Algorithms

Most of the existing optimization algorithms are SI based [2]. SI-based algorithms mimic the collective behavior of multiple individuals who communicate and react in nature, e.g., swarms, herds, schools, or flocks of creatures [1, 5, 18] that seem to be unintelligent at first. However, due to their self-organization over the course of generations or iterations following a set of rules that simulate their behavioral movements and interactions [19], they can do some sort of collective intelligence for finding the global optimum. Metaheuristic algorithms are classified according to the main dominant classes: SI-based algorithms and Evolutionary algorithms [5]. Root causes of SI popularity are: (a) simplicity of implementation compared to the evolutionary algorithms, which have extra operators like selection, crossover, etc., (b) efficiency, due to sharing information among individuals (agents), self-organization [2], and the fact that it is easy to parallelize individuals, and (c) agents' capabilities for learning and co-evolution over the course of iterations.

A list of selected SI algorithms from 2014 up to now is presented in Table 1.1.

1.3.2 BI-not-SI-Based Algorithms

Most recent studies about classifying the optimization algorithms, classify algorithm into main categories or dominant classes like the SI based algorithms [5]. In fact, most of the natural inspired algorithms are bio-inspired [2]. Such as shown above in Figure 1.1, SI is a subset of BI, and BI is a subset of NI.

Not all BI algorithms are SI-based. For example, GA does not depend on swam behavior but on some operators like crossover and mutation. Thus, this

TABLE 1.1: A list of SI algorithms from 2014 up to now

Index	Algorithm	Reference	Year	Main Feature
1	Artificial Butterfly	[4]	2017	Exploration and exploitation balancing
2	Lion Optimizer	[20]	2016	Based on lions behavior (e.g., prey capturing ... etc.)
3	Sperm Whale	[21]	2016	Bases on worst and best solutions in exploitation phase
4	Whale Optimizer	[22]	2016	Humpback whale behavior (e.g., searching prey ... etc.)
5	Elephant Herding	[23]	2015	Clan updating operator and separating operator
6	Ions Motion	[24]	2015	Mimics attraction and repulsion of anions and cations
7	Moth-Flame	[25]	2015	Targets constrained problems with unknown search spaces
8	Vortex Search	[26]	2015	Single based solution - performs numerical function optimization
9	Water Wave	[27]	2015	For global optimization problems
10	Chaotic Bat	[28]	2014	Echolocation bat behavior. Increases global search mobility

category is called BI-not-SI algorithms. However, the differential evolution (DE) [29] is similar to the GA; they are classified as a type evolutionary algorithms [5], but DE is not easy to be classified because it does not mimic any biological behavior, but it is tentatively considered as a BI algorithm. Some relevant algorithms are summarized in Table 1.2.

TABLE 1.2: A list of selected BI-based algorithms

Index	Algorithm	Reference	Year	Main Feature
1	Human Group (HGO)	[30]	2016	Decision making in groups
2	Symbiotic Organisms Search	[31]	2014	Numerical and engineering design problems
3	Dolphin Echolocation	[32]	2013	Echolocation (navigation and hunting)
4	Egyptian Vulture	[33]	2013	Egyptian Vultures for acquiring food - combinatorial optimization problems
5	Mine Blast (MBA)	[34]	2013	Mine bomb explosion concept - constrained engineering optimization problems

1.3.3 Natural Science-Based Algorithms

As mentioned earlier, the subset relation between the BI algorithms and NI algorithms means that there is a set of algorithms that are not BI but NI depending on the source of inspiration [13]. These algorithms mimic certain physical phenomena or chemical law (e.g., gravity, ions motion, electrical charges, river systems, etc.). Thus we call them physical inspired and chemical inspired, respectively (i.e., physical-based and chemical-based algorithms). It is desirable to group resulting algorithms from these types to the same category [2], so we called them natural science-based algorithms; where some of their fundamental laws are similar. Some selected algorithms from 2012 up to now are shown in Table 1.3.

TABLE 1.3: A list of some recent natural science-based algorithms

Index	Algorithm	Reference	Year	Main Feature
1	Ions Motion	[24]	2015	Anions and cations behavior
2	Gases Brownian Motion	[35]	2013	Features of gas molecules for solving various functions
3	Magnetic Charged System	[36]	2013	Magnetic and electrical forces based Biot–Savart law
4	Ray Optimization	[37]	2012	Snell's light refraction law
5	Chemical Reaction Optimization	[38]	2012	Chemical reactions - solve discrete and continuous problems

1.3.4 Natural Phenomena-Based Algorithms

The main feature in classifying the metaheuristic algorithms is the source of inspiration [2, 13]. As discussed above, a subset of the NI algorithms is called BI and the other subset encompasses the physical-based and chemical-based algorithms [39]. But still there is a subset from the natural inspired algorithms that does not belong to any of the previous subsets. This subset of the remaining algorithms mimics various characteristics (e.g., social and emotional characteristics) from different sources [2]. The new subset is called natural phenomena-based algorithms (NP-based), some of them are listed in Table 1.4.

TABLE 1.4: A list of natural phenomena-based algorithms

Index	Algorithm	Reference	Year	Main Feature
1	Virus Colony Search	[40]	2016	Global numerical and engineering optimization problems
2	Crow Search	[41]	2016	Different engineering design problems
3	Artificial Cooperative Search	[42]	2013	Numerical optimization problems
4	Backtracking Optimization Search	[43]	2013	Multi-modal and real-valued numerical optimization problems
5	Differential Search	[44]	2012	From geocentric Cartesian to geodetic

1.4 Variants of Nature-Inspired Algorithms

This section proposes different variants for the NI algorithms. Algorithms could be categorized into eight main categories depending on problem type: continuous, discrete, parallel, distributed, binary, chaotic, multi-objective and hybrid algorithms. In fact, in each category from the previous one, at least two variants have been proposed to solve either feature selection or optimization problems. It is clear that the main factor categories rely on is the actual perspective; however the main focus of the proposed attempt in this research work is the solution type. Well-known variants that this section focuses on are binary, chaotic, multi-objective and hybrid algorithms.

1.4.1 Binary Algorithms

Some variants for binary NI algorithms have been proposed and depicted in Table 1.5.

1.4.2 Chaotic Algorithms

This section presents some variants for chaotic NI algorithms depicted in Table 1.6.

TABLE 1.5: A list of binary NI algorithms

Index	Algorithm	Reference	Year	Main Feature
1	S-shaped Binary Whale	[45]	2019	Feature selection
2	Binary Whale	[46]	2017	Feature selection
3	Binary Salp	[47]	2017	Feature selection
4	Binary Dragonfly	[48]	2016	Feature selection
5	Binary Clonal Flower Pollination	[49]	2016	Feature selection
6	Binary Ant-lion	[50]	2016	Feature selection
7	Binary Cockroach Swarm	[51]	2016	Function optimization
8	Binary Flower Pollination	[52]	2016	Feature selection
9	Binary Grey Wolf	[53]	2016	Feature selection
10	Binary Cat Swarm	[54]	2015	Discrete optimization

TABLE 1.6: A list of chaotic NI algorithms

Index	Algorithm	Reference	Year	Main Feature
1	Chaotic Bird Swarm	[55]	2018	Parameter optimization
2	Chaotic Ant-lion	[56]	2017	Parameter optimization
3	Chaotic Grey Wolf	[57]	2017	Optimization
4	Chaos-embedded Krill Herd	[58]	2016	Volt ampere reactive (VAR) dispatch - power system
5	Chaos-enhanced Cuckoo Search	[59]	2016	Enhances basic cuckoo search
6	Chaotic Ant-lion	[60]	2016	Feature selection
7	Chaotic Brain Storm	[61]	2015	Optimization
8	Chaotic Bat Swarm	[62]	2015	Optimization
9	Chaotic Krill Herd	[63]	2014	Enhances convergence acceleration in the Krill Herd
10	Chaotic binary PSO	[64]	2011	Feature selection

1.4.3 Multi-objective Algorithms

In this section, some variants have been proposed for multi-objective NI algorithms depicted in Table 1.7.

TABLE 1.7: A list of multi-objective nature-inspired algorithms

Index	Algorithm	Reference	Year
1	Multi-objective Sine-Cosine	[68]	2019
2	Multi-objective Modified Flower Pollination	[65]	2018
3	Multi-objective Grasshopper	[66]	2018
4	Multi-objective Grey Wolf	[67]	2017
5	Multi-objective Ant-lion	[69]	2017
6	Multi-objective Moth Flame	[70]	2016
7	Multi-objective Cuckoo Search	[71]	2016
8	Multi-objective PSO	[10]	2015
9	Multi-objective Flower	[72]	2013
10	Multi-objective Bat	[73]	2011

1.4.4 Hybrid Algorithms

Sometimes nature-inspired algorithms produce inaccurate solutions for some real-time optimization problems. Due to recommendations on local optima convergence, convergence speed, and large-scale spaces coverage, dynamic problems, constraint and multiple objectives problems remain as hot research topics. Imitating the best feature in nature to solve optimization problems is an ongoing work. Hybridizing algorithms require a huge number of functions to be evaluated, which recommends the future research work to be concentrated on agents' stagnation problems to improve algorithms performance and increase results accuracy. Some variants have been proposed for hybrid nature-inspired algorithms in Table 1.8.

TABLE 1.8: A list of hybrid nature-inspired algorithms

Index	Algorithm	Reference	Year	Main Feature
1	Grasshopper & Opposition-based Learning	[74]	2018	Optimization problems
2	Grasshopper & Support Vector Machines (SVMs)	[75]	2018	Bio-signal problems
3	Elephant Herding & Support Vector Regression (SVR)	[76]	2018	Bio-signal problems
4	Hybrid Swarm	[77]	2018	Bio-signal and biomedical problems
5	Grey Wolf enhanced SVMs	[78]	2017	Bio-signal problems
6	Water Wave Optimization & SVMs	[79]	2017	Bio-signal problems
7	Whale Optimizer & SVR	[80]	2017	Energy problems
8	PSO & SVR	[81]	2017	Energy problems
9	Whale Optimization & SVR	[82]	2017	Energy problems
10	Dimer Swarm & PSO	[83]	2017	Global optimization
11	Whale Optimization & Simulated Annealing	[84]	2017	Feature selection
12	Krill Herd-based Cuckoo Search	[85]	2016	Engineering optimization and global optimization
13	PSO & Gravitational Search	[86]	2016	Global optimization
14	Flower Pollination & Bee Pollinator	[87]	2016	Cluster analysis
15	Firefly & PSO	[88]	2016	Bundle branch block detection

1.5 A Review of the Most Recent NI Algorithms

This section discusses the most recent NI algorithms clarifying their main inspiration source, motivation scenario, mathematical model, and heuristics. They include: Artificial Butterfly Optimization (ABO) [4], Grasshopper Optimization Algorithm (GOA) [9], Salp Swarm Optimization Algorithm (SSA) [5], Spotted Hyena Optimization Algorithm (SHO) [39], and Chemotherapy Science Optimization Algorithm (CSA) [1].

1.5.1 Artificial Butterfly Optimization Algorithm

- **Inspiration:** This algorithm is a BI algorithm, which mimics the mate-finding strategy of woodland butterfly (speckled woods). There are two main strategies of the male butterflies: (1) dominated butterflies, which are called sunspot butterflies; they monopolize large sunspots where the more females visit (60% of the male butterflies), and (2) sub-dominated butterflies, which are called canopy butterflies (40% of the male butterflies).

- **Mathematical Model:** Three types of butterfly flights are called: (1) sunspot, (2) canopy, and (3) free flight. In the first two modes, butterflies change their locations toward the most visited regions by females (i.e., bigger sunspot regions). The ith butterfly location (X_i^{t+1}) is determined by Eq. 1.1:

$$X_i^{t+1} = X_i^t + \frac{X_k^t - X_i^t}{||X_k^t - X_i^t||}.(Ub - Lb).step.rand() \tag{1.1}$$

Where k: random neighbor butterfly, $k \neq i$, and t: iteration number. Function $rand$ generates random values $\in [0 : 1]$. Ub and Lb are the upper and lower bounds, respectively. The operator $step$ is the flight distance, which decreases gradually from 1 to $step_e$ in order to neutralize exploration and exploitation following Eq. 1.2.

$$step = 1 - (1 - step_e).\frac{E}{maxE} \tag{1.2}$$

Where E: number of current evaluations, and $maxE$: maximum number of evaluations.
The third one is the free flight mode, where the new butterfly position is identified according to Eq. 1.3.

$$X_i^{t+1} = X_i^t - 2.a.rand() - a.D \tag{1.3}$$

$$D = |2.rand().X_i^t - X_k^t| \tag{1.4}$$

Where the operator a decreases from 2 to 0 after a round of iterations.

- **Pseudo-Code:** Figure 1.2 shows butterfly pseudo-code.
- **Heuristics:** By implementing different flights and comparing results with the other algorithms (Artificial Bee Colony (ABC), PSO, Chaotic PSO, and GA). The authors recommended the algorithm to reduce computation time.

1.5.2 Grasshopper Optimization Algorithm

- **Inspiration:** This algorithm mimics the natural behavior of the grasshopper such as “how grasshoppers migrate over large distances”. There are two possible strategies: (1) when grasshoppers become nymph size, they move like a rolling cylinder, (2) when they become adult size, they move abruptly forming a swarm in the air seeking food source.

```
1. Initialize the locations of butterfly population
2. Evaluate the fitness of every butterfly
3. While not meet the terminal condition
      3.1 Sort all butterflies by their fitness
      3.2 Select butterflies with better fitness to form sunspot, the rest form canopy butterflies.
      3.3 For each sunspot butterfly
            a. fly to one new location according to sunspot flight mode
            b. Evaluate the fitness of the new sunspot
            c. apply greedy selection on the original location and the new one
      3.4 End For
      3.5 For each canopy butterfly
            a. Fly to one randomly selected sunspot butterfly according to canopy flight mode.
            b. Evaluate the fitness
            c. If better fitness:
                  o apply greedy selection on the original location and the new one
            d. Else
                  o Fly to new location according to free flight mode
      3.6 End For
4. End While
Post process results
```

FIGURE 1.2: Butterfly optimizer pseudo-code [4]

- **Mathematical Model:** The grasshoppers change their positions (X_i) following a set of conditions (e.g., gravity, wind advection, and social forces), such as shown in Eq. 1.5.

$$X_i = r_1.S_i + r_2.G_i + r_3.A_i \tag{1.5}$$

Where r_1, r_2, and r_3 are random values (to generate random behavior for the ith grasshopper), Operators S_i, G_i, and A_i represent social interaction, gravity force, and wind advection, respectively.

The social force could be determined by Eq. 1.6. Types of social forces are the attraction force (to exploit promising regions) and the repulsion force (to explore search regions). When attraction and repulsion do not exist, grasshoppers become in what is called a $comfortzone$. Strength of social forces (S component in Eq. 1.5) could be determined by Eq. 1.7.

$$s(r) = fe^{\frac{-r}{l}} - e^{-r} \tag{1.6}$$

Where l : attractive length scale and f : the intensity of attraction.

$$S_i = \sum_{j=1\ j\neq i}^{N} s(d_{ij})\widehat{d_{ij}} \tag{1.7}$$

where $d_{ij} = |X_j - X_i|$ (distance between grasshoppers) and their unity vector $\widehat{d_{ij}} = \frac{X_j - X_i}{d_{ij}}$. Because Nymph flights are affected by the wind, which will be the direction of the best solution $\widehat{T_d}$ such as indicated in the following equation. After substitution, the resulting equation will be as follows (Eq. 1.8)

$$X_i^d = c\left(\sum_{j=1\ j\neq i}^{N} c\frac{ub_d - lb_d}{2} s(|X_j^d - X_i^d|)\frac{X_j - X_i}{d_ij} \right) + \widehat{T_d} \tag{1.8}$$

where d is the dimension. Coefficient c reduces $comfortzone$ over course of iterations and balances between local and global search. The coefficient c is determined as follows:

$$c = c_{max} - l\frac{c_{max} - c_{min}}{L} \tag{1.9}$$

Where l: current iteration, L: maximum number of iterations.

- **Pseudo-Code:** the Grasshopper algorithm is shown in Figure 1.3.

1. Initialize the swarm $\mathbf{X_i}$ (i=0,1, …, n)
2. Initialize **cmax, cmin**, and **maxiterations**
3. Calculate the fitness of each search agent
4. **T**= the best search agent
5. While (l< maxiterations)
    5.1 update the coefficient **(c)**
    5.2 for each search agent
        5.2.1 normalize the distance between the grasshoppers in [1,4]
        5.2.2 update current search agent position (X_i^d)
        5.2.3 bring current search agent back if it goes outside the boundaries
    5.3 end for
6. End while
7. Return **T**

FIGURE 1.3: Pseudo-code of the Grasshopper optimizer [9]

- **Heuristics:** GOA can enhance the average fitness by improving the initial random population and targeting the global optimum over round of iterations. This algorithm targets problems with unknown search spaces.

1.5.3 Salp Swarm Optimization Algorithm

- **Inspiration:** Salp swarm optimization algorithm (SSA) mimics salps' behavior in nature (navigating and foraging). Multi-objective of SSA is called MSSA. Salps are a member in the Salpidae family. They form a swarm called *salpchain*. Groups of salp population: (a) leaders (first salp in the front of each chain) and (b) followers (the rest of the salps of the chain).

- **Mathematical Model:** SSA search space is determined by n-dimensions (i.e., n-variable) like other swarm optimization algorithms. The leader salp search for food source (FS) and remaining salps come after the leader gradually over a number of iterations. The leading salp changes its position (X_j^1) in dimension j according to the following equation:

$$X_j^i = \begin{cases} FS_j + c_1(c_2(ub_j - lb_j) + lb_j) & c_3 \geq 0 \\ FS_j - c_1(c_2(ub_j - lb_j) + lb_j) & c_3 < 0 \end{cases} \tag{1.10}$$

Where $c1 = 2e^{-(\frac{4l}{L})^2}$ and $c2$ and $c3$: random numbers. Follower salps change their positions according to the following equation:

$$X_j^i = \frac{1}{2}(X_j^i - X_j^{i-1}) \tag{1.11}$$

- **Pseudo-Code:** SSA and MSSA.
MSSA is a new version of SSA that handles multi-objective problems by implementing a repository for storing multiple solutions in SSA in order to compare solutions at each step. If a salp of the current population dominated a set of salps in the repository, then it is added to the repository and they are removed, and it is discarded otherwise. SSA and MSSA pseudo-codes are shown in Figure 1.4.

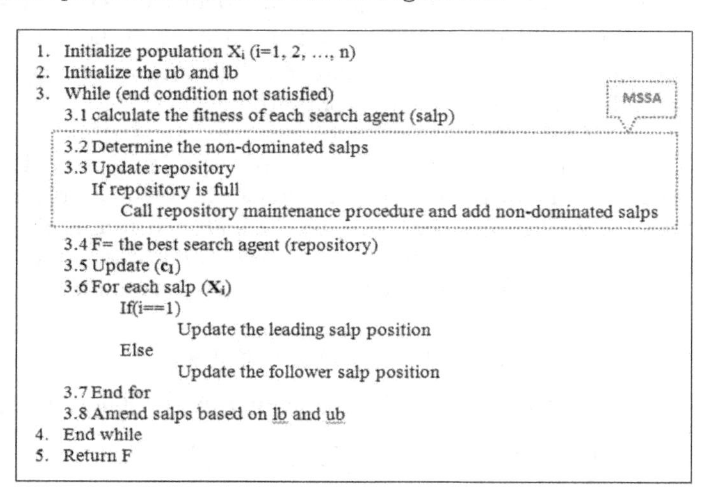

FIGURE 1.4: Salp optimization pseudo-code [5]

- **Heuristics:** The main features of the SSA optimizer are as follows:
  - Easy to implement.
  - Keep/store best solutions.

- The coeffecient c1 is a controlling parameter, which decreases adaptively to allow exploring the search space then exploiting it.
- Balances between the exploration and exploitation for avoiding entrapment of the local optima.

1.5.4 Spotted Hyena Optimization Algorithm

- **Inspiration:** This algorithm mimics spotted hyenas' behavior in the nature (searching, encircling, and attacking the prey).
- **Mathematical Model:**
  - *Encircling:* Search agents try to encircle their prey by moving toward the prey changing their positions. Eq. 1.12 is used to know how far away the spotted hyena is from the prey, while Eq. 1.13 is used to define new positions of the spotted hyenas.

$$\vec{\mathrm{D}}_h = |\vec{\mathrm{A}}.\vec{\mathrm{P}}_p - \vec{\mathrm{P}}(Y)| \tag{1.12}$$

$$\vec{\mathrm{P}}(Y+1) = \vec{\mathrm{P}}_p(Y) - \vec{\mathrm{E}}.\vec{\mathrm{D}}_h \tag{1.13}$$

Where $\vec{\mathrm{P}}$: spotted hyena, $\vec{\mathrm{P}}_p(Y)$: the prey at iteration Y, and $\vec{\mathrm{D}}_h$: distance between $\vec{\mathrm{P}}$ and $\vec{\mathrm{P}}_p(Y)$. The coefficient vector $\vec{\mathrm{A}} = 2.r\vec{\mathrm{d}}_1$ and the coefficient vector $\vec{\mathrm{E}} = 2\vec{\mathrm{h}}.r\vec{\mathrm{d}}_2 - \vec{\mathrm{h}}$, where $\vec{\mathrm{d}}_1$ and $\vec{\mathrm{d}}_2$ are random vectors in [0,1], and $\vec{\mathrm{h}} = 5.(1 - iteration/max_{iteration})$ and decreases step by step from 5 to 0.

  - *Hunting:* In this stage, the best search agent knows the location of the prey, and other search agents make clusters and groups towards it, saving the obtained best solutions after a round of iterations. Distance between the first best solution and other spotted hyenas could be defined similarly to Eq. 1.12, such as follows:

$$\vec{\mathrm{D}}_h = |\vec{\mathrm{B}}.\vec{\mathrm{P}}_h - \vec{\mathrm{P}}_a| \tag{1.14}$$

and the cluster $\vec{\mathrm{C}}_h$ could be evaluated by the following equation:

$$\vec{\mathrm{C}}_h = \vec{\mathrm{P}}_a + \vec{\mathrm{P}}_{a+1} + \vec{\mathrm{P}}_{a+2} + \ldots + \vec{\mathrm{P}}_{a+N} \tag{1.15}$$

where N: total count of spotted hyenas.

  - *Searching and Attacking:* According to the value of the vector $\vec{\mathrm{E}}$ spotted hyenas determine when to search (if $|\vec{\mathrm{E}}| \leq 1$) and when to attack (if $|\vec{\mathrm{E}}| > 1$). Spotted hyenas updated their positions during the attack stage as follows:

$$\vec{\mathrm{P}}(X+1) = \frac{\vec{\mathrm{C}}_h}{N} \tag{1.16}$$

- **Pseudo-Code:**
  Pseudo-code of spotted hyena optimizer is shown in Figure 1.5.

```
1. Input: the spotted hyenas population P_i(i = 1, 2, ..., n)
2. Output: the best search agent:
3. procedure SHO:
4. Initialize the parameters h, B, E, and N
5. Calculate the fitness of each search agent
6. P_h = the best search agent
7. C_h = the group or cluster of all far optimal solutions
8. while (x < Max number of iterations) do
       for each search agent do
              Update the position of current agent by Eq. (17)
       end for
       Update h, B, E, and N
       Check if any search agent goes beyond the given search space and then adjust it
       Calculate the fitness of each search agent
       Update P_h if there is a better solution than previous optimal solution
       Update the group C_h w.r.t P_h
       x = x + 1
9. end while
10. return P_h
11. end procedure
```

FIGURE 1.5: Pseudo-code of spotted hyena optimizer [39]

- **Heuristics:** Spotted hyena optimizer is used to solve real-life optimization problems with minimal efforts to find optimal solutions.

1.5.5 Chemotherapy Science Optimization Algorithm

- **Inspiration:** This algorithm mimics the chemotherapy method for treating cancer patients, which aims to limit and decrease number of infeasible cells. This algorithm simulates the treatment cycle, e.g., treatment doses represent number of iterations, and the reset period between two sequential treatment doses in order to recover cells is the period in which the algorithm generates a number of solutions to start the next iteration.

- **Mathematical Model:**
  - *Limiting the initial cell position element bounds (CPE):* Phase of determining tumor size starts by determining and tuning lb and ub of the cell position elements.
  - *Initial cell generation:* There are four approaches for determining the initial cell:
    * Limiting bounds in the previous step
    * Implementing relaxation methods (linear programming)
    * Implementing greedy methods (problem-based)

* Composition of multiple approaches (e.g., random search, variable bounds, etc.).

After that, propose $X^{p_o} = (X_1^{p_o}, X_2^{p_o}, ..., X_J^{p_o})$, which means the pth solution at the iteration 0 for cells from $1 : J$ over the population p.

- *Evaluating health status and tumor position:* Searching for new healthy cells in the tumor means generating new solutions in the infeasible space. Resulting pairs of cells in CSA from the search process are compared based on the (a) Tumor Position (TP) and Tumor Size (TS). Eq. 1.17 calculates TP_i^{pt} for constraint i for all cells at the iteration t, while Eq. 1.18 calculates the aggregation of all TP_i^{pt} at all constraints, where V_i is the constraint weight.

$$TP_i^{pt} = \begin{cases} \sum_{j \in J} a_{ij} X_j^{pt} - b_i \sum_{j \in J} a_{ij} X_j^{pt} & \geq b_i \\ 0 & O.W \end{cases} \quad \forall i \in I \tag{1.17}$$

$$TTP^{pt} = \sum_{i \in I} TP_i^{pt} V_i \tag{1.18}$$

the total value of tumor size of all cells at iteration t is calculated using Eq. 1.19.

$$TS^{pt} = \sum_{j \in J} C_j X_j^{pt} \tag{1.19}$$

the arithmetic mean of all tumor positions and tumor sizes over all iterations is evaluated by Eq. 1.20 and Eq. 1.21.

$$ATS^t = \frac{1}{|CG(t)|} \sum_{p \in CG(t)} TS^{pt} \tag{1.20}$$

$$ATTP^t = \frac{1}{|CG(t)|} \sum_{p \in CG(t)} TTP^{pt} \tag{1.21}$$

where $|CG(t)|$: cardinality of first ranked cells group at iteration t $(\in [1 : T])$.

- *Searching neighborhood cells*: By generating random cells and determining cell area (CA) in tumor
- *Converting cancerous cells to healthy ones.*

- **Pseudo-Code:** Figure 1.6 shows the general steps of the chemotherapy optimizer.

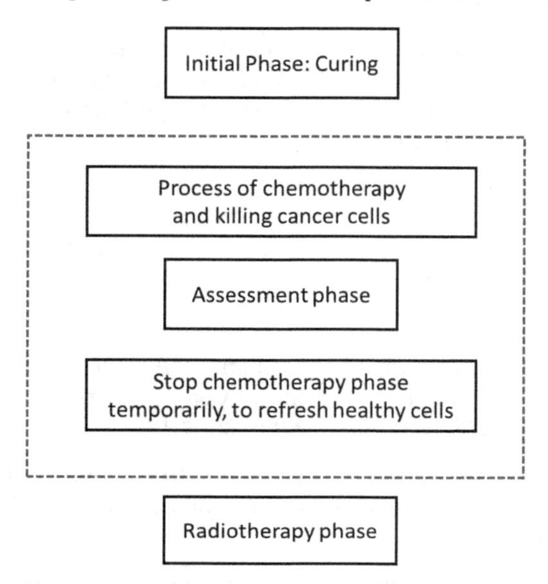

FIGURE 1.6: Pseudo-code of chemotherapy optimizer algorithm [1]

- **Heuristics:**
  - This algorithm explores cells far from the tumor border to its outside targeting to convert infected cells to appropriate healthy cells by comparing each pair based on the TTP and TS criteria.
  - Greedy methods could be implemented to generate initial population and to convert infected cells.

1.6 Conclusion

Current research trends in the metaheuristic algorithms focus on identifying new ideas for solving complex problems, and also on inventing new advancement of algorithms for solving new problems or for enhancing performance of existing algorithms. Our aim in this review is to guide researchers to gain better knowledge about latest metaheuristic algorithms which cover a large-scale from the real world problems. Different sources of inspiration produce different types of metaheuristic algorithms. This review briefly summarizes all natural-inspired algorithms according to the sources of inspiration. They could be categorized into: swarm-intelligence-based algorithms (SI), bio-inspired algorithms (BI), nature science-based algorithms (NS), and natural phenomena-based algorithms (NPA). It is worth pointing out that there is no unique classification for the metaheuristic algorithms, and this try is just for the purpose of information only.

Bibliography

[1] M. H. Salmani and K. Eshghi, "A metaheuristic algorithm based on chemotherapy science: Csa," *Journal of Optimization, https://doi.org/10.1155/2017/3082024*, 2017.

[2] I. Fister Jr, X.-S. Yang, I. Fister, J. Brest, and D. Fister, "A brief review of nature-inspired algorithms for optimization," *arXiv preprint arXiv:1307.4186*, 2013.

[3] S. Binitha, S. S. Sathya *et al.*, "A survey of bio-inspired optimization algorithms," *International Journal of Soft Computing and Engineering*, vol. 2, no. 2, pp. 137–151, 2012.

[4] X. Qi, Y. Zhu, and H. Zhang, "A new meta-heuristic butterfly-inspired algorithm," *Journal of Computational Science*, vol. 23, pp. 226–239, 2017.

[5] S. Mirjalili, A. H. Gandomi, S. Z. Mirjalili, S. Saremi, H. Faris, and S. M. Mirjalili, "Salp swarm algorithm: a bio-inspired optimizer for engineering design problems," *Advances in Engineering Software*, vol. 114, pp. 163–191, 2017.

[6] K. K. Vardhini and T. Sitamahalakshmi, "A review on nature-based swarm intelligence optimization techniques and its current research directions," *Indian Journal of Science and Technology*, vol. 9, no. 10, 2016.

[7] B. K. Panigrahi, Y. Shi, and M.-H. Lim, *Handbook of Swarm Intelligence: Concepts, Principles and Applications.* Springer Science & Business Media, 2011, vol. 8.

[8] J. Kennedy and R. Eberhart, "Particle swarm optimization," in *Proc. IEEE Int. Conf. Neural Networks*, vol. 4. IEEE Service Center, Piscataway, NJ, 1995, pp. 1941–1948.

[9] S. Saremi, S. Mirjalili, and A. Lewis, "Grasshopper optimisation algorithm: Theory and application," *Advances in Engineering Software*, vol. 105, pp. 30–47, 2017.

[10] C. Dai, Y. Wang, and M. Ye, "A new multi-objective particle swarm optimization algorithm based on decomposition," *Information Sciences*, vol. 325, pp. 541–557, 2015.

[11] Z. Beheshti, S. M. Shamsuddin, and S. S. Yuhaniz, "Binary accelerated particle swarm algorithm (bapsa) for discrete optimization problems," *Journal of Global Optimization*, vol. 57, no. 2, pp. 549–573, 2013.

[12] X. Teng, H. Dong, and X. Zhou, "Adaptive feature selection using v-shaped binary particle swarm optimization," *PloS One*, vol. 12, no. 3, p. e0173907, 2017.

[13] H. Zang, S. Zhang, and K. Hapeshi, "A review of nature-inspired algorithms," *Journal of Bionic Engineering*, vol. 7, pp. S232–S237, 2010.

[14] S. Kirkpatrick, C. D. Gelatt, and M. P. Vecchi, "Optimization by simulated annealing," *Science*, vol. 220, no. 4598, pp. 671–680, 1983.

[15] X.-S. Yang, "Firefly algorithm, stochastic test functions and design optimisation," *International Journal of Bio-Inspired Computation*, vol. 2, no. 2, pp. 78–84, 2010.

[16] J. H. Holland, "Genetic algorithms," *Scientific American*, vol. 267, no. 1, pp. 66–73, 1992.

[17] X.-S. Yang and S. Deb, "Cuckoo search via lévy flights," in *Nature & Biologically Inspired Computing, 2009. NaBIC 2009. World Congress on.* IEEE, 2009, pp. 210–214.

[18] I. Boussaïd, J. Lepagnot, and P. Siarry, "A survey on optimization metaheuristics," *Information Sciences*, vol. 237, pp. 82–117, 2013.

[19] M. N. Ab Wahab, S. Nefti-Meziani, and A. Atyabi, "A comprehensive review of swarm optimization algorithms," *PLoS One*, vol. 10, no. 5, p. e0122827, 2015.

[20] M. Yazdani and F. Jolai, "Lion optimization algorithm (loa): a nature-inspired metaheuristic algorithm," *Journal of Computational Design and Engineering*, vol. 3, no. 1, pp. 24–36, 2016.

[21] A. Ebrahimi and E. Khamehchi, "Sperm whale algorithm: An effective metaheuristic algorithm for production optimization problems," *Journal of Natural Gas Science and Engineering*, vol. 29, pp. 211–222, 2016.

[22] S. Mirjalili and A. Lewis, "The whale optimization algorithm," *Advances in Engineering Software*, vol. 95, pp. 51–67, 2016.

[23] G.-G. Wang, S. Deb, and L. d. S. Coelho, "Elephant herding optimization," in *Computational and Business Intelligence (ISCBI), 2015 3rd International Symposium on.* IEEE, 2015, pp. 1–5.

[24] B. Javidy, A. Hatamlou, and S. Mirjalili, "Ions motion algorithm for solving optimization problems," *Applied Soft Computing*, vol. 32, pp. 72–79, 2015.

[25] S. Mirjalili, "Moth-flame optimization algorithm: A novel nature-inspired heuristic paradigm," *Knowledge-Based Systems*, vol. 89, pp. 228–249, 2015.

[26] B. Doğan and T. Ölmez, "A new metaheuristic for numerical function optimization: Vortex search algorithm," *Information Sciences*, vol. 293, pp. 125–145, 2015.

[27] Y.-J. Zheng, "Water wave optimization: a new nature-inspired metaheuristic," *Computers & Operations Research*, vol. 55, pp. 1–11, 2015.

[28] A. H. Gandomi and X.-S. Yang, "Chaotic bat algorithm," *Journal of Computational Science*, vol. 5, no. 2, pp. 224–232, 2014.

[29] R. Storn and K. Price, "Differential evolution–a simple and efficient heuristic for global optimization over continuous spaces," *Journal of Global Optimization*, vol. 11, no. 4, pp. 341–359, 1997.

[30] I. De Vincenzo, I. Giannoccaro, and G. Carbone, "The human group optimizer (hgo): mimicking the collective intelligence of human groups as an optimization tool for combinatorial problems," *arXiv preprint arXiv:1608.01495*, 2016.

[31] M.-Y. Cheng and D. Prayogo, "Symbiotic organisms search: a new metaheuristic optimization algorithm," *Computers & Structures*, vol. 139, pp. 98–112, 2014.

[32] A. Kaveh and N. Farhoudi, "A new optimization method: dolphin echolocation," *Advances in Engineering Software*, vol. 59, pp. 53–70, 2013.

[33] C. Sur, S. Sharma, and A. Shukla, "Egyptian vulture optimization algorithm–a new nature inspired meta-heuristics for knapsack problem," in *The 9th International Conference on Computing and InformationTechnology (IC2IT2013)*. Springer, 2013, pp. 227–237.

[34] A. Sadollah, A. Bahreininejad, H. Eskandar, and M. Hamdi, "Mine blast algorithm: A new population based algorithm for solving constrained engineering optimization problems," *Applied Soft Computing*, vol. 13, no. 5, pp. 2592–2612, 2013.

[35] M. Abdechiri, M. R. Meybodi, and H. Bahrami, "Gases brownian motion optimization: an algorithm for optimization (gbmo)," *Applied Soft Computing*, vol. 13, no. 5, pp. 2932–2946, 2013.

[36] A. Kaveh, M. A. M. Share, and M. Moslehi, "Magnetic charged system search: a new meta-heuristic algorithm for optimization," *Acta Mechanica*, vol. 224, no. 1, pp. 85–107, 2013.

[37] A. Kaveh and M. Khayatazad, "A new meta-heuristic method: ray optimization," *Computers & Structures*, vol. 112, pp. 283–294, 2012.

[38] A. Y. Lam and V. O. Li, "Chemical reaction optimization: A tutorial," *Memetic Computing*, vol. 4, no. 1, pp. 3–17, 2012.

[39] G. Dhiman and V. Kumar, "Spotted hyena optimizer: A novel bio-inspired based metaheuristic technique for engineering applications," *Advances in Engineering Software*, vol. 114, pp. 48–70, 2017.

[40] M. D. Li, H. Zhao, X. W. Weng, and T. Han, "A novel nature-inspired algorithm for optimization: Virus colony search," *Advances in Engineering Software*, vol. 92, pp. 65–88, 2016.

[41] A. Askarzadeh, "A novel metaheuristic method for solving constrained engineering optimization problems: crow search algorithm," *Computers & Structures*, vol. 169, pp. 1–12, 2016.

[42] P. Civicioglu, "Artificial cooperative search algorithm for numerical optimization problems," *Information Sciences*, vol. 229, pp. 58–76, 2013.

[43] K. Guney, A. Durmus, and S. Basbug, "Backtracking search optimization algorithm for synthesis of concentric circular antenna arrays," *International Journal of Antennas and Propagation*, vol. 2014, pp. 11, 2014.

[44] P. Civicioglu, "Transforming geocentric cartesian coordinates to geodetic coordinates by using differential search algorithm," *Computers & Geosciences*, vol. 46, pp. 229–247, 2012.

[45] A. G. Hussien, A. E. Hassanien, E. H. Houssein, S. Bhattacharyya, and M. Amin, "S-shaped binary whale optimization algorithm for feature selection," in *Recent Trends in Signal and Image Processing*. Springer, 2019, pp. 79–87.

[46] A. G. Hussien, E. H. Houssein, and A. E. Hassanien, "A binary whale optimization algorithm with hyperbolic tangent fitness function for feature selection," in *Intelligent Computing and Information Systems (ICICIS), 2017 Eighth International Conference on*. IEEE, 2017, pp. 166–172.

[47] A. G. Hussien, A. E. Hassanien, and E. H. Houssein, "Swarming behaviour of salps algorithm for predicting chemical compound activities," in *Intelligent Computing and Information Systems (ICICIS), 2017 Eighth International Conference on*. IEEE, 2017, pp. 315–320.

[48] S. Mirjalili, "Dragonfly algorithm: a new meta-heuristic optimization technique for solving single-objective, discrete, and multi-objective problems," *Neural Computing and Applications*, vol. 27, no. 4, pp. 1053–1073, 2016.

[49] S. A.-F. Sayed, E. Nabil, and A. Badr, "A binary clonal flower pollination algorithm for feature selection," *Pattern Recognition Letters*, vol. 77, pp. 21–27, 2016.

[50] E. Emary, H. M. Zawbaa, and A. E. Hassanien, "Binary ant lion approaches for feature selection," *Neurocomputing*, vol. 213, pp. 54–65, 2016.

[51] I. C. Obagbuwa and A. P. Abidoye, "Binary cockroach swarm optimization for combinatorial optimization problem," *Algorithms*, vol. 9, no. 3, p. 59, 2016.

[52] Z. A. E. M. Dahi, C. Mezioud, and A. Draa, "On the efficiency of the binary flower pollination algorithm: application on the antenna positioning problem," *Applied Soft Computing*, vol. 47, pp. 395–414, 2016.

[53] E. Emary, H. M. Zawbaa, and A. E. Hassanien, "Binary grey wolf optimization approaches for feature selection," *Neurocomputing*, vol. 172, pp. 371–381, 2016.

[54] B. Crawford, R. Soto, N. Berríos, F. Johnson, F. Paredes, C. Castro, and E. Norero, "A binary cat swarm optimization algorithm for the non-unicost set covering problem," *Mathematical Problems in Engineering*, vol. 2015, pp. 8, 2015.

[55] F. H. Ismail, E. H. Houssein, and A. E. Hassanien, "Chaotic bird swarm optimization algorithm," in *International Conference on Advanced Intelligent Systems and Informatics.* Springer, 2018, pp. 294–303.

[56] A. Tharwat and A. E. Hassanien, "Chaotic antlion algorithm for parameter optimization of support vector machine," *Applied Intelligence*, vol. 48, no. 3, pp. 670–686, 2018.

[57] M. Kohli and S. Arora, "Chaotic grey wolf optimization algorithm for constrained optimization problems," *Journal of Computational Design and Engineering*, 2017.

[58] A. Mukherjee and V. Mukherjee, "Chaos embedded krill herd algorithm for optimal var dispatch problem of power system," *International Journal of Electrical Power & Energy Systems*, vol. 82, pp. 37–48, 2016.

[59] L. Huang, S. Ding, S. Yu, J. Wang, and K. Lu, "Chaos-enhanced cuckoo search optimization algorithms for global optimization," *Applied Mathematical Modelling*, vol. 40, no. 5-6, pp. 3860–3875, 2016.

[60] H. M. Zawbaa, E. Emary, and C. Grosan, "Feature selection via chaotic antlion optimization," *PloS one*, vol. 11, no. 3, p. e0150652, 2016.

[61] Z. Yang and Y. Shi, "Brain storm optimization with chaotic operation," in *Advanced Computational Intelligence (ICACI), 2015 Seventh International Conference on.* IEEE, 2015, pp. 111–115.

[62] A. R. Jordehi, "Chaotic bat swarm optimisation (cbso)," *Applied Soft Computing*, vol. 26, pp. 523–530, 2015.

[63] S. Saremi, S. M. Mirjalili, and S. Mirjalili, "Chaotic krill herd optimization algorithm," *Procedia Technology*, vol. 12, pp. 180–185, 2014.

[64] L.-Y. Chuang, C.-H. Yang, and J.-C. Li, "Chaotic maps based on binary particle swarm optimization for feature selection," *Applied Soft Computing*, vol. 11, no. 1, pp. 239–248, 2011.

[65] E. H. Houssein and M. Abd-ELfattah, "Multi-objective modified flower pollination algorithm for maximizing lifetime in wireless sensor networks," *International Journal of Systems Signal Control and Engineering Application*, vol. 11, no. 1, pp. 20–29, 2018.

[66] A. Tharwat, E. H. Houssein, M. M. Ahmed, A. E. Hassanien, and T. Gabel, "Mogoa algorithm for constrained and unconstrained multi-objective optimization problems," *Applied Intelligence*, vol. 48, no. 8, pp. 2268–2283, 2018.

[67] A. Sahoo and S. Chandra, "Multi-objective grey wolf optimizer for improved cervix lesion classification," *Applied Soft Computing*, vol. 52, pp. 64–80, 2017.

[68] M. A. Tawhid and V. Savsani, "Multi-objective sine-cosine algorithm (mo-sca) for multi-objective engineering design problems," *Neural Computing and Applications*, vol. 31, Supplement 2, pp. 915–929, 2019.

[69] S. Mirjalili, P. Jangir, and S. Saremi, "Multi-objective ant lion optimizer: a multi-objective optimization algorithm for solving engineering problems," *Applied Intelligence*, vol. 46, no. 1, pp. 79–95, 2017.

[70] S. J. Nanda *et al.*, "Multi-objective moth flame optimization," in *Advances in Computing, Communications and Informatics (ICACCI), 2016 International Conference on.* IEEE, 2016, pp. 2470–2476.

[71] W. Yamany, N. El-Bendary, A. E. Hassanien, and E. Emary, "Multi-objective cuckoo search optimization for dimensionality reduction," *Procedia Computer Science*, vol. 96, pp. 207–215, 2016.

[72] X.-S. Yang, M. Karamanoglu, and X. He, "Multi-objective flower algorithm for optimization," *Procedia Computer Science*, vol. 18, pp. 861–868, 2013.

[73] X.-S. Yang, "Bat algorithm for multi-objective optimisation," *International Journal of Bio-Inspired Computation*, vol. 3, no. 5, pp. 267–274, 2011.

[74] A. A. Ewees, M. A. Elaziz, and E. H. Houssein, "Improved grasshopper optimization algorithm using opposition-based learning," *Expert Systems with Applications*, vol. 112, pp. 156–172, 2018.

[75] A. Hamad, E. H. Houssein, A. E. Hassanien, and A. A. Fahmy, "Hybrid grasshopper optimization algorithm and support vector machines for automatic seizure detection in eeg signals," in *International Conference on Advanced Machine Learning Technologies and Applications*. Springer, 2018, pp. 82–91.

[76] A. E. Hassanien, M. Kilany, E. H. Houssein, and H. AlQaheri, "Intelligent human emotion recognition based on elephant herding optimization tuned support vector regression," *Biomedical Signal Processing and Control*, vol. 45, pp. 182–191, 2018.

[77] E. H. Houssein, A. A. Ewees, and M. A. ElAziz, "Improving twin support vector machine based on hybrid swarm optimizer for heartbeat classification," *Pattern Recognition and Image Analysis*, vol. 28, no. 2, pp. 243–253, 2018.

[78] A. Hamad, E. H. Houssein, A. E. Hassanien, and A. A. Fahmy, "A hybrid eeg signals classification approach based on grey wolf optimizer enhanced svms for epileptic detection," in *International Conference on Advanced Intelligent Systems and Informatics*. Springer, 2017, pp. 108–117.

[79] M. Kilany, E. H. Houssein, A. E. Hassanien, and A. Badr, "Hybrid water wave optimization and support vector machine to improve emg signal classification for neurogenic disorders," in *Computer Engineering and Systems (ICCES), 2017 12th International Conference on*. IEEE, 2017, pp. 686–691.

[80] S. Osama, A. Darwish, E. H. Houssein, A. E. Hassanien, A. A. Fahmy, and A. Mahrous, "Long-term wind speed prediction based on optimized support vector regression," in *Intelligent Computing and Information Systems (ICICIS), 2017 Eighth International Conference on*. IEEE, 2017, pp. 191–196.

[81] E. H. Houssein, "Particle swarm optimization-enhanced twin support vector regression for wind speed forecasting," *Journal of Intelligent Systems*, 2017.

[82] S. Osama, E. H. Houssein, A. E. Hassanien, and A. A. Fahmy, "Forecast of wind speed based on whale optimization algorithm," in *Proceedings of the 1st International Conference on Internet of Things and Machine Learning*. ACM, 2017, p. 62.

[83] J. Yang and M. P. Waller, "A hybrid dimer swarm optimizer," *Computational and Theoretical Chemistry*, vol. 1102, pp. 98–104, 2017.

[84] M. M. Mafarja and S. Mirjalili, "Hybrid whale optimization algorithm with simulated annealing for feature selection," *Neurocomputing*, vol. 260, pp. 302–312, 2017.

[85] G.-G. Wang, A. H. Gandomi, X.-S. Yang, and A. H. Alavi, "A new hybrid method based on krill herd and cuckoo search for global optimisation tasks," *International Journal of Bio-Inspired Computation*, vol. 8, no. 5, pp. 286–299, 2016.

[86] A. Yadav, K. Deep, J. H. Kim, and A. K. Nagar, "Gravitational swarm optimizer for global optimization," *Swarm and Evolutionary Computation*, vol. 31, pp. 64–89, 2016.

[87] R. Wang, Y. Zhou, S. Qiao, and K. Huang, "Flower pollination algorithm with bee pollinator for cluster analysis," *Information Processing Letters*, vol. 116, no. 1, pp. 1–14, 2016.

[88] P. Kora and K. S. R. Krishna, "Hybrid firefly and particle swarm optimization algorithm for the detection of bundle branch block," *International Journal of the Cardiovascular Academy*, vol. 2, no. 1, pp. 44–48, 2016.

Chapter 2

Hybrid Cartesian Genetic Programming Algorithms: A Review

Johnathan Melo Neto
Federal University of Juiz de Fora, Brazil
E-mail: jmmn.mg@gmail.com

Heder S. Bernardino
Federal University of Juiz de Fora, Brazil
E-mail: heder@ice.ufjf.br

Helio J.C. Barbosa
National Laboratory for Scientific Computing, Brazil
E-mail: hcbm@lncc.br

2.1 Introduction 28
2.2 Metaheuristics 30
2.2.1 Single-Solution Methods 31
2.2.2 Population-Based Methods 31
2.2.2.1 Evolution strategies 31
2.2.2.2 Differential evolution 32
2.2.2.3 Biogeography-based optimization 32
2.2.2.4 Non-dominated sorting genetic algorithm . 33
2.2.2.5 Harmony search 34
2.2.2.6 Estimation of distribution algorithms 35
2.2.2.7 Ant colony optimization 35
2.2.2.8 Particle swarm optimization 36
2.3 Fundamentals of Cartesian Genetic Programming 37
2.3.1 Historical Context 37
2.3.2 Encoding 37
2.3.3 Evolution Scheme 38
2.3.4 Parameters 39
2.3.5 Advantages and Drawbacks 39
2.4 Literature Review on Hybrid Metaheuristics 40
2.5 Hybrid Cartesian Genetic Programming Algorithms 42
2.5.1 Motivation 42
2.5.1.1 CGP combined with ant colony optimization 42

2.5.1.2 CGP combined with biogeography-based optimization and opposition-based learning 44
2.5.1.3 CGP combined with differential evolution . 45
2.5.1.4 CGP combined with estimation of distribution algorithm 47
2.5.1.5 CGP combined with NSGA-II 48
2.5.1.6 CGP combined with harmony search 49
2.5.1.7 CGP combined with particle swarm optimization 51
2.6 Discussion on Hybrid CGP Algorithms 52
2.7 Future Directions of Hybrid CGP Algorithms 52
2.8 Concluding Remarks 55
Acknowledgment 55
Bibliography 56

2.1 Introduction

The generation of computer programs utilizing Genetic Programming (GP) techniques has drawn the interest of many scientists after the publication of Koza's book [41]. Since then, researchers have been developing new ideas to improve GP's performance and expanding its applicability. Some of those GP applications include discrete-time optimal control problems [50], financial trading, bioinformatics, and many others [62]. In this direction, Cartesian Genetic Programming (CGP), primarily proposed in [51], is a form of GP in which a computer program is described as a directed-indexed-acyclic graph. CGP represents a node grid addressed by a Cartesian coordinate system. Its genotype is constituted by function, connection, and output genes. The function genes represent the nodes' functionalities by utilizing a look-up table, the connection genes define the nodes' inputs, and the output genes define the program's outputs. In its primary version [54], CGP uses the $(1+\lambda)$ Evolution Strategy (ES) [65].

Although the initial applications using CGP have focused on digital circuitry designs [53, 55], researchers later applied the method to many other domains. A brief description of some applications is presented.

Clarke [8] explored how CGP can be applied to find solutions to Control Engineering applications. The work focused on the design of controllers by applying CGP to evolve control strategies, given a specification set. That approach was named Cartesian Genetic Control (CGC) and was compared to classical design solutions in two typical Control Engineering problems. CGC achieved the desired specifications, demonstrating faster response than the comparative approach, and showing a promising alternative for controller design.

Hara et al. [29] applied CGP to a multi agent control model, in which imitated elements are determined by the roulette-type selection based on their collaboration to team performance. That model was compared to three distinct methods: homogeneous, heterogeneous, and conventional CGP. The experimental results confirmed the effectiveness of the proposal, which selects a promising agent and succeeds in generating better offspring for the next iterations.

CGP has also been used for image processing. As such, the CGP-based image processing technique developed by Harding et al. [31] is an automatic generator of programs, which uses a subset of the OpenCV library functionalities. That approach was applied to robotics, noise reduction, and medical imaging, demonstrating high competitiveness and generating models that incorporate advanced domain knowledge.

In the area of hearing aid systems, Ullah [78] presented a sign recognition method that uses CGP to recognize the American Sign Language (ASL) gestures. The dataset contained 26 ASL gestures representing the letters of the English alphabet. The experimental findings were relevant, with high average accuracy. There are many other significant applications using CGP, and a more complete list can be found in [52, 67].

The main benefits of CGP include the neutral genetic drift, the absence of program bloat, the adequacy to Multiple-Input Multiple-Output (MIMO) problems, and the reusability of internally calculated values [77].

Despite all the existing benefits, the exclusive use of CGP may lead to suboptimal results for certain problems. A higher performance can be achieved when a combination of CGP with another search technique is implemented. For instance, the graph-based representation scheme of CGP can be well-suited to represent a given model; however, when additional numerical parameters must also be optimized, the use of an auxiliary search method is a suitable alternative. In this scenario, the joint use of two (or more) methods can enhance the performance of the solutions, instead of solving the problem using a single technique.

For instance, machine learning methods are widely used in the generation of predictive models, and a potential application of CGP is the evolution of Artificial Neural Networks (ANNs) [39], given its representation (using a graph). In this case, CGP is suitable to obtain the topology of the ANN, but it is not appropriate to optimize its weights, as CGP is designed to evolve solutions in a discrete search space. Thus, a better-suited search technique may be used to assist CGP in finding more accurate parameters for the ANNs [49].

Although CGP is widely applied in the literature, just a few works have investigated the hybridization of CGPs with other search methods. To address the lack of works in that direction, this chapter provides a critical review of CGP algorithms that make use of hybrid approaches. Additionally, a detailed description of CGP and suggestions on how it can be combined with other search mechanisms will be presented. Although GP hybrids have been developed in the literature [58, 63], the main purpose of this chapter is the review of hybrid Cartesian GP algorithms. Also, to the best of our knowledge, no such review is available in the literature.

Some works have followed the strategy of developing hybrid versions of CGP and have achieved interesting results. In those works, CGP is combined with the following techniques: Ant Colony Optimization [30, 43], Estimation of Distribution Algorithm [9], Particle Swarm Optimization [59], Harmony Search [21], Differential Evolution [49], Biogeography-based Optimization [83], and Non-dominated Sorting Genetic Algorithm [35]. All of those hybrids will be presented in detail in this chapter.

This review aims at presenting the current hybrid CGP algorithms and stimulating the development of new algorithms of this class. We believe that the structural and functional flexibility of CGP, combined with different search techniques, should be further investigated and applied to different domains, potentially leading to new relevant findings.

The present work is arranged as follows. Section 2.2 provides the basic notions of metaheuristics and brief descriptions of some algorithms. Section 2.3 describes the implementation and characteristics of CGP in detail. Section 2.4 presents an overview of some significant hybridization techniques in the literature. Section 2.5 provides a complete review of the hybrid CGP algorithms and the main motivation for their development. An in-depth discussion on the hybrid CGP algorithms is presented in Section 2.6. Potential future research directions related to hybrid CGP is given in Section 2.7. Some conclusions can be found in Section 2.8.

2.2 Metaheuristics

Metaheuristics are self-learning methods originated from the field of intelligent systems that arise in biology, physics and other knowledge domains [48]. According to [44], the term "metaheuristic" describes a major field called Stochastic Optimization, which is characterized by the existence of stochasticity in the optimization scheme, as either Monte Carlo randomness in the search process, noise in measurements, or both. Those algorithms produce near-optimal solutions and offer a good compromise between the solutions' quality and their computational costs.

This section presents a summary of the metaheuristics used for optimization and which are hybridized with CGP in proposals from the literature. Those metaheuristics are organized here in two major groups: (1) Single-Solution Methods and (2) Population-Based Methods. Given the relevance of CGP here, this chapter devotes a separate section to its complete description.

2.2.1 Single-Solution Methods

Single-solution methods are metaheuristics that maintain only one candidate solution at a time. They provide different strategies by which a solution can be: (i) refined by neighborhood search techniques, making small, bounded, and random changes, or (ii) improved by global search techniques, making large random changes.

Many single-solution methods can be found in the literature. Some of them are Random Search [4], Stochastic Hill Climbing [22], Iterated Local Search [73], and Tabu Search [26].

2.2.2 Population-Based Methods

Population-based methods keep a group of solutions instead of just one [44]. Most of those algorithms are nature-inspired, a well-known group of techniques termed as Evolutionary Algorithms (EAs). EAs belong to the Evolutionary Computation field, which concerns the techniques inspired by the Darwinian evolution process. Those algorithms correspond to simplified variants of the evolution mechanisms by using biological concepts such as natural selection, reproduction, genetic recombination, selective pressure, and environmental adaptability. The main concept is that a limited amount of individuals is able to reproduce, as they meet a selection requirement, and the next generations will move toward those candidates [80]. Some examples of Evolutionary Algorithms are Evolution Strategies [65] and Genetic Algorithms [33].

Another popular class of techniques is known as Swarm Algorithms. They are computational models created from the cooperation and intelligence of uniform agents [5]. Such cooperative behavior is distributed and self-organized. The information is typically stored in all agents or transmitted through the environment. The main Swarm Algorithms are Particle Swarm Optimization [38] and Ant Colony Optimization [19]. Given the importance of the population-based methods in this chapter, some of the algorithms from that group are detailed next.

2.2.2.1 Evolution strategies

Evolution Strategies (ES) [65] are inspired by the species-level process of evolution. ES evolves the solutions using adaptive operators [5]. In general, the ES operates on a population of μ parents and λ children. Initially, μ parents are randomly generated ($1 \leq \mu \leq \lambda$). Then, λ children are produced by mutation (and possibly recombination) of the μ parents picked using uniform probability. In the (μ, λ)-ES variant, the top μ individuals amid the λ children are selected for the coming iteration, and the other ones are eliminated. In the $(\mu{+}\lambda)$-ES version, the top μ individuals among all the $(\mu{+}\lambda)$ individuals of the population are picked for the next generation. The selection and reproduction processes are reiterated until some predefined stop rule is reached.

2.2.2.2 Differential evolution

The Differential Evolution (DE) [72] algorithm has been designed for function optimization in R^n [5]. The individual is denoted by a real numerical vector. The algorithm involves the maintenance of a population subject to variation, evaluation, and selection. A mutation creates a candidate based on the weighted difference between two individuals added to a third one. In the primary version of DE, denoted by DE/rand/1/bin, the so-called donor vector v_i is created by

$$v_i^{(j+1)} = x_{r_3}^{(j)} + F \cdot (x_{r_1}^{(j)} - x_{r_2}^{(j)}), \tag{2.1}$$

where *Pop* is the population size, j is the iteration counter, $i = 1, \ldots, Pop$, F is a predefined parameter, and $x_{r_1}^{(j)}$, $x_{r_2}^{(j)}$ and $x_{r_3}^{(j)}$ are three vectors randomly chosen from the population ($i \neq r_1 \neq r_2 \neq r_3$). CR is the crossover rate, a parameter used to create a trial vector $u_i^{(j+1)}$ by performing a combination between the target vector $x_i^{(j)}$ and the donor vector $v_i^{(j+1)}$ as

$$u_{ik}^{(j+1)} = \begin{cases} v_{ik}^{(j+1)}, & \text{if } r_4 \leq CR \text{ or } r_{id} = k \\ x_{ik}^{(j)}, & \text{otherwise,} \end{cases} \tag{2.2}$$

where r_4 is a random number in $[0, 1)$, $k = 1, \ldots, n$; r_{id} is a random integer from $\{1, \ldots, n\}$, and n is the vector dimension. After, the fitness of the newly created candidate solution is compared to the fitness of the target vector. More information about DE can be found in [13].

2.2.2.3 Biogeography-based optimization

Biogeography-Based Optimization (BBO) [69] is an evolutive algorithm that has the science of biogeography as inspiration. The individuals, also called habitats, represent candidate solutions and they are composed of solution features. Those features are called Suitability Index Variables (SIVs), and each individual's fitness is described by its corresponding Habitat Suitability Index (HSI). Therefore, high-HSI solutions are good individuals, whereas the low-HSI individuals represent the poor ones. The migration and mutation are the main BBO operators.

BBO shares the features between the individuals by applying the migration operator exchanging characteristics between high-HSI and low-HSI individuals. Each individual possesses its emigration rate μ and immigration rate λ, which represents the probability that an individual is selected as a habitat of emigration or immigration, respectively. For each individual, the rates' values are set based on its fitness, as indicated by

$$\lambda_i = I \cdot \left(1 - \frac{k(i)}{n_{pop}}\right), \tag{2.3}$$

$$\mu_i = E \cdot \frac{k(i)}{n_{pop}}, \tag{2.4}$$

where n_{pop} is the amount of candidates in the population, and $k(i)$ is the rank of the i-th candidate solution in a sorted list based on the population fitness. The parameters I and E are the max values of immigration and emigration rates, respectively. Therefore, a good individual relatively has a low λ and a high μ. Once those rates are calculated, the migration scheme can be represented as $S_{i,SIV} \leftarrow S_{j,SIV}$, where S_i is the immigratant individual that presents an immigration rate λ_i, and S_j is the emigrant individual that presents an emigration rate μ_j. The migration equation indicates that a characteristic of S_i is substituted by a characteristic of S_j.

The mutation operator is applied in order to transform the solution features. A random SIV in the i-th individual S_i is substituted by a random SIV, according to a given mutation rate. The mutation rate is determined by

$$m_i = m_{max} \cdot \left(\frac{1 - P_i}{P_{max}} \right), \tag{2.5}$$

where m_{max} is the predefined max mutation rate, $P_{max} = \arg\max P_i$, $i = 1, ..., n_{pop}$, and P_i is the solution probability of the i-th candidate, which is calculated using λ_i and μ_i. More details of BBO can be found in [69].

2.2.2.4 Non-dominated sorting genetic algorithm

Genetic Algorithms (GAs) [33, 47] are evolutionary systems inspired by the Darwinian selection mechanism, which is a biological process where the fitter individuals are probably the winners in a competitive environment. The numerical parameters that define an individual are encoded in a data structure called a chromosome. The fitness reflects the suitability of the chromosome to resolve the task, and it is based on the objective function of the optimization task.

In more complex problems, it is common to observe multiple and conflicting objectives. Those problems are called Multiobjective Problems (MOPs), and their goal is optimizing k objective functions simultaneously. Formally, an MOP is defined as minimizing (or maximizing) $F(\boldsymbol{x}) = (f_1(\boldsymbol{x}), ..., f_k(\boldsymbol{x}))$ subject to $g_i(\boldsymbol{x}) \leq 0$, $i = \{1, ..., m\}$, and $h_j(\boldsymbol{x}) = 0$, $j = \{1, ..., p\}$. The MOP solution minimizes (or maximizes) the vector components calculated using $F(\boldsymbol{x})$ where $\boldsymbol{x}$ is a n-dimensional decision variable vector from some universe Ω. The expressions $g_i(\boldsymbol{x}) \leq 0$ and $h_j(\boldsymbol{x}) = 0$ represent constraints that need to be satisfied, and Ω presents all possible $\boldsymbol{x}$ [11].

In the context of MOPs, the notions of Pareto front and Pareto dominance are relevant. Assuming minimization, a vector $\boldsymbol{u} = (u_1, ..., u_k)$ dominates another vector $\boldsymbol{v} = (v_1, ..., v_k)$ (denoted by $\boldsymbol{u} \preceq \boldsymbol{v}$) if and only if $u_i \leq v_i \, \forall i \in \{1, ..., k\}$, and $\exists i \in \{1, ..., k\} : u_i < v_i$. Thus, the Pareto optimal set, $\mathcal{P}^*$, is defined as $\mathcal{P}^* := \{\boldsymbol{x} \in \Omega \,|\, \neg\exists\, \boldsymbol{x}' \in \Omega;\ F(\boldsymbol{x}') \preceq F(\boldsymbol{x})\}$ and the Pareto front $\mathcal{PF}^*$ can be defined as $\mathcal{PF}^* := \{\boldsymbol{u} = F(\boldsymbol{x}) \,|\, \boldsymbol{x} \in \mathcal{P}^*\}$.

As GA operates with a set of candidate solutions, special mechanisms are necessary in order to solve MOPs. However, auxiliary approaches are necessary. In this direction, Non-dominated Sorting Genetic Algorithm (NSGA) [71] is a multiobjective optimization GA that optimizes a candidate solution's set so that it approximates the Pareto front of the problem. The individuals are sorted into sub-populations (ranks) based on the Pareto dominance (non-dominated sorting). The non-dominated solutions are grouped into one rank with a dummy fitness value. The individuals of this rank share their dummy fitness values, so the population diversity is maintained. Then this rank is skipped, and the method processes the next layer of non-dominated solutions. This iteration is repeated until the classification of all solutions. To maintain the diversity of the population, the solutions share their dummy fitness values. Also, a proportionate random selection is used, so the first-rank individuals are replicated more frequently compared to the other individuals. This selection generates better Pareto front search procedures, which results in the convergence of the population in direction to these locations. Also, the sharing scheme aids the population distribution on the Pareto front of the optimization task.

Deb et al. [14] later extended NSGA to a new approach called NSGA-II, which ranks and sorts each solution with respect to the non-dominance level. Afterward, it executes evolutionary operations to build a new offspring set, which combines both parents and offspring before splitting the new combined set into fronts. Next, NSGA-II uses a crowding distance in order to prioritize individuals in less populated regions in the objectives space. This scheme keeps the diversification and helps the fitness landscape's exploration [11].

2.2.2.5 Harmony search

Harmony Search (HS) [25] is a population-based method that mimics the behavior of an orchestra while trying to compose a harmonic melody. It is considered a physical algorithm due to the inspiration from a physical process. In the metaphor of HS: each instrument corresponds to a variable (feature) in an individual, each pitch of an instrument represents a given value of a variable, and the pitch range denotes the decision variable limits [5]. A harmony corresponds to a candidate solution, i.e., the combination of pitches of different instruments, and the harmony aesthetic appreciation represents the objective of the problem. HS seeks a good harmony over time through small variations of the pitches of the instruments, which results in an upgrade over the objective function. The four HS steps are described hereafter. Step 1: Initialize a Harmony Memory (HM). Step 2: Improvise a novel harmony from HM. Step 3: Include the new harmony in HM, if the novel harmony is better than one of the HM harmonies (and remove the worst one). Step 4: If stopping conditions are not met, go to Step 2.

HM is a group of harmonies, i.e., a group of candidate solutions. HM initializes with random values, and is updated whenever any of the newer individuals is better than one of the individuals currently stored in HM.

To create a new harmony, HS uses a parameter known as Harmony Memory Considering Rate (HMCR), ranging in the interval [0,1]. If a randomly generated value is smaller than HMCR, then HS generates pitches within the possible pitch range, without using the HM. Otherwise, HS reuses the pitches that are present in HM.

To escape local optima, HS uses the Pitch Adjusting Rate (PAR). The PAR imitates the pitch settings of the instruments. That mechanism shifts the pitches to neighboring values within a given interval [25]. For instance, if there are five possible values such as $\{2, 3, 6, 9, 8\}$, $\{6\}$ can be shifted to $\{3\}$ or $\{9\}$ in the pitch adjusting scheme. A PAR of 0.20 denotes that the technique picks a neighboring upper value with 10% probability or a neighboring lower value with 10%.

Despite the promising results, HS has been criticized for being a particular case of the Evolution Strategies method [82]. Implementation details of HS can be found in [5, 24].

2.2.2.6 Estimation of distribution algorithms

Probabilistic Algorithms perform the search space exploration using a probabilistic model of candidate solutions [5]. Most of those methods are called Estimation of Distribution Algorithms (EDAs), and pertain to the evolutionary algorithms class that replaces the variation operators by obtaining a probabilistic model of picked individuals, and sampling it to create novel individuals.

EDAs generate a random initial population of solutions over the search space within a uniform distribution [60]. The basic EDAs consider that individuals are denoted by fixed-length arrays over a limited set of symbols. At each generation, good individuals are picked from the population. Then, a probabilistic model of the picked individuals is constructed, and that new probabilistic model is sampled to create newer individuals. That estimate is obtained by using the frequency of utilization of components in promising individuals. After, the newer individuals are included in the population, substituting some of the old individuals (or all of them). The scheme repeats until it reaches the stop rules.

Also, the EDAs can be applied as an iterative search scheme for the construction of a probabilistic model that creates global optima [60]. Firstly, the model codifies a uniform distribution over all possible solutions. In each generation, the model is updated to build better individuals. An effective EDA is supposed to generate a model that finds all global optima. Some examples of Probabilistic Algorithms are the Bayesian Optimization Algorithm [61] and the Univariate Marginal Distribution Algorithm [57].

2.2.2.7 Ant colony optimization

Ant Colony Optimization (ACO) [19] is a swarm-type algorithm based on the ants' cooperativism. In the ACO metaphor, an ant is oriented by pheromone levels on the disposable paths, where ants can move from source to the final destination. Initially, ants move randomly around the environment.

If an ant finds a path that generates a good solution, it then begins to lay down pheromone on such a path. If that path is frequently utilized by many ants, more pheromone is stored on it. Thus, there is a higher likelihood that an ant entering that path will move to a promising future path to find a better solution. Likewise, there is a low probability of an ant picking a path that presents a low pheromone level. Therefore, positive feedback attracts more ants to good paths, which are later refined through constant utilization [5].

In this direction, the purpose of ACO is to explore heuristic and historical data to create individuals and store into the history the information learned during the search. That history is placed in a pheromone table, which contains the pheromone levels which are used by the ant in constructing its path. The pheromone table is updated at each iteration, and that update scheme depends on the ACO variant used.

The decision rule of the ants is defined by [18]: (1) the values of a local data structure known as ant-routing table, which can be acquired by a combination of local pheromone trails and heuristic values, (2) the ant's memory, that stores the history, and (3) the problem's constraints.

The main ACO variants are Ant System [15], Ant Colony System [17], Max-Min Ant System [74], and Rank-Based Ant System [6].

2.2.2.8 Particle swarm optimization

Particle Swarm Optimization (PSO) [38] is a metaheuristic paradigm that has the movement of fish and birds as inspiration. Metaphorically, the swarm agents move through an environment following the fitter elements, and frequently biasing their motion in direction to historically promising areas of the environment [5]. PSO has a population (swarm) of solutions (particles) that move in the search space (environment) and cooperate.

The algorithm is performed by improving the position of the particles using their velocities, the current best global position, and the current best position of each particle. The velocity has a huge impact on the swarm behavior and is defined by

$$\vec{v}_i(t+1) \leftarrow w \cdot \vec{v}_i(t) + c_{soc} \cdot R_1(t) \cdot [\,\vec{g}(t) - \vec{\gamma}_i(t)\,] + c_{cog} \cdot R_2(t) \cdot [\,\vec{b}_i(t) - \vec{\gamma}_i(t)\,], \quad (2.6)$$

where c_{soc} is the social factor, c_{cog} is the cognitive factor, $\vec{b}_i$ is the current best position of the i-th particle, $\vec{b}_i$ is the best position identified by the population, w is a user-defined parameter, and R_1 and R_2 are diagonal matrices with random elements, to avoid a deterministic movement of the particles.

After the velocity update, the particle position is also updated as

$$\vec{\gamma}_i(t+1) \leftarrow \vec{\gamma}_i(t) + \vec{v}_i(t+1), \quad (2.7)$$

where the i-th particle updates its new position $\vec{\gamma}_i(t+1)$ by considering its current location $\vec{\gamma}_i(t)$ and velocity $\vec{v}_i(t+1)$ at iteration t. The execution of PSO is finalized when a stop criterion is satisfied. Further implementation details of PSO can be found in [38, 84].

2.3 Fundamentals of Cartesian Genetic Programming

After a brief review of some metaheuristics, a full comprehension of another one, named CGP, is essential to investigate how it can be further improved by using the hybrid formulations. Therefore, this section presents the fundamentals of CGP and its characteristics.

2.3.1 Historical Context

Genetic Programming (GP) [41, 62] is a special form of Genetic Algorithm, developed for automatic problem solving, that, from a high-level description of the requirements, builds programs without explicit programming of the solution structure by the user. Since 1958, several works based on this principle have been proposed [12, 23, 70]. However, the technique only spread after the publication of Koza [41] in 1992.

In general, GP evolves a set of programs that, along with the generations, stochastically create better programs compared to those from previous generations. The quality of a program depends on its performance towards the problem of interest. In their basic form, GP programs are usually represented as syntactic trees. In advanced forms, programs are composed of multiple components (e.g., sub-routines), represented as a set of trees clustered under a root node [62]. To create new programs, the structure of the sub-trees and their nodes are modified by the use of the genetic operators of mutation and crossover. More details on GP can be found in [41, 62].

Later in 1997, Miller et al. [55] proposed a new form of GP to design digital circuits by evolving the functions and connections of a grid of logic gates. That new form of GP was nominated Cartesian Genetic Programming (CGP) in [51], as it uses a node grid addressed in a Cartesian coordinate system.

2.3.2 Encoding

CGP programs represent directed-acyclic graphs. The CGP nodes are originally organized in n_r rows and n_c columns, forming a rectangular two-dimensional grid of $n_r \times n_c$ nodes, as shown in Figure 2.1. Each node can receive its entries from nodes in the former columns, or from inputs. The parameter called levels-back determines the number of preceding columns which can connect to a node in the present column. For example, if levels-back is 2, a node can only receive as input the values from the nodes located in the two columns immediately anterior to the present column. It should be noted that this parameter does not prevent the nodes from connecting to program inputs.

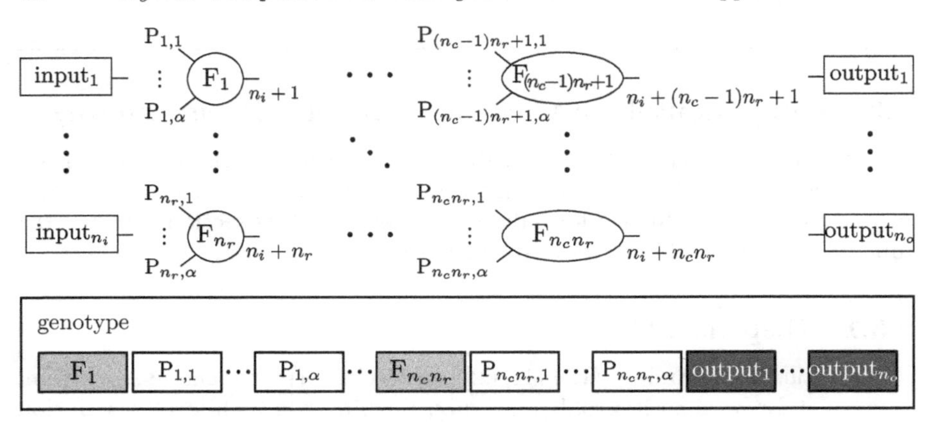

FIGURE 2.1: Representation of CGP using a grid of nodes with n_r rows, n_c columns, n_i inputs, n_o outputs, and α entries per node. The corresponding genotype is illustrated below the graph.

However, the two-dimensional grid topology is unnecessarily restrictive, as a more general configuration can be obtained by arranging the nodes in a single row, with the quantity of columns equals to the quantity of nodes and the levels-back equals to the quantity of columns. Thus, each node is able to connect with any node on its left and with the program inputs. An example of this non-restrictive configuration is shown in Figure 2.2.

The CGP genotype is composed of function, connection, and output genes. Function genes represent the functionality of the nodes by utilizing a look-up table to match the code to its function, e.g., the code 0 corresponds to the addition operation. Connection genes determine the nodes' inputs. Output genes define the program outputs. The nodes that are not used for the calculation of at least one output are called inactive, and therefore do not need to be computed during the execution of the graph.

2.3.3 Evolution Scheme

CGP uses the Evolution Strategies $(\mu + \lambda)$-ES coupled with point-type or probabilistic-type mutation. The method begins with a set of μ randomly generated candidate solutions. These candidates are evaluated and each of those μ candidates produces λ/μ children by using mutation. Hence, λ new individuals are created. The next iteration has μ parents and λ new children. CGP continues its execution until a stop rule is met. Typically, that rule corresponds to a max number of generations permited.

The mutation is executed by assigning random valid values for the selected genes. In the point mutation, the user previously establishes the amount of genes that must be mutated, thus the genes are randomly selected up to the

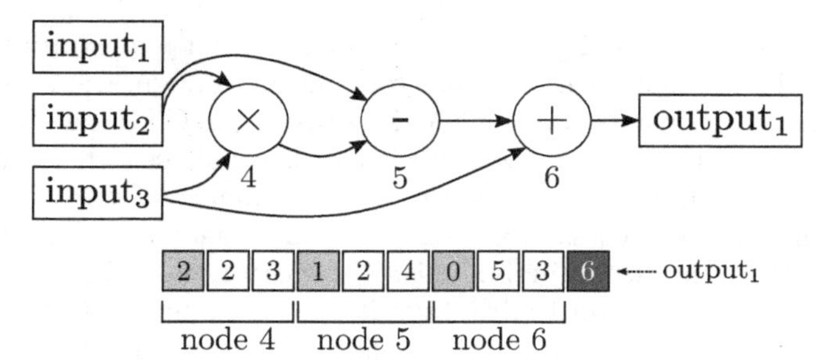

FIGURE 2.2: Illustration of a non-restrictive configuration CGP graph, presenting 3 inputs ($n_i = 3$), 3 nodes (where $n_r = 1$ and $n_c = 3$) and 1 output ($n_o = 1$). There is no inactive node. The genotype is given below the graph. The gray values denote the transfer functions of the nodes according to a look-up table (0: addition, 1: subtraction, and 2: multiplication), the white values denote the connection genes of each node, and the final value (dark gray) represents the output connection.

predefined maximum amount. On the other hand, in the probabilistic mutation, each gene can mutate with a given probability. Note that the probabilistic mutation allows a more flexible search in the solution space [77] as mutation can occur in any gene, in some genes, or in all genes, whereas the point mutation defines the exact amount of genes to be mutated, restricting the search for different individuals.

2.3.4 Parameters

The parameters of CGP control the evolutionary process, the topology of the graphs, and the mathematical operations available to the nodes. Therefore, their choice is essential to the proper functioning of the algorithm. To execute CGP in its non-restrictive configuration (using a single row and several columns), the following parameters must be defined: (1) max amount of nodes ($n_r \times n_c$), (2) max amount of entries per node, (3) function set available to the nodes, (4) number of parents to generate the children (μ), (5) number of children (λ), (6) mutation rate, which in the case of the point mutation determines the proportion of genes to be mutated, and in the case of the probabilistic mutation determines the probability of each gene to be mutated, and (7) stop criteria.

2.3.5 Advantages and Drawbacks

Many advantages of CGP may explain its widespread use in the literature. Some of them, described in [77], are highlighted here. The first advantage of CGP is the reusability of internally calculated values. In this direction,

CGP also presents the ability to handle Multiple-Input and Multiple-Output (MIMO) problems due to its flexible topological configuration. As a wide amount of applications often involve MIMO tasks, the CGP may be an alternative to be considered when dealing with such problems. The second characteristic allows individuals to enhance the performance of the solutions over the iterations without necessarily resulting in an exaggerated and disproportionate growth in the size of individuals. That feature is commonly known as resilience to bloat. The third advantage is the capability of CGP to escape from local minima due to mutations in inactive nodes. When the nodes of an individual remain inactive after the mutations have been applied, the genotypes are modified but the phenotypes are maintained. Therefore, although those solutions travel through new locations of the search space, such displacements do not alter their fitness values. Later, if such nodes become active after new mutations, there will be solutions with different phenotypes among them, making the escape from local optima more likely, as the search space is better explored for different solutions. That property is commonly known in the literature as neutral genetic drift.

In general, the literature does not explicitly present many drawbacks of CGP. First, traditional CGP makes use of mutation as the only genetic operator, which means that it does not use any type of crossover to exchange genetic information between individuals. Therefore, according to [83] the exploitation of CGP can be compromised, i.e., it can lack the ability to search for new regions and potentially find a better local optimum. Second, CGP is designed for discrete optimization problems [52], so the method may somehow yield below-ideal results when applied to continuous problems.

2.4 Literature Review on Hybrid Metaheuristics

Up to this point, we have described some important metaheuristics that can be found in the literature. Now, we advance our discussion by introducing some hybridization techniques that include metaheuristics in their formulations.

In general, hybrid systems are based on the complementary perspective of search algorithms [34], since they can cooperate and complement their benefits to potentially improve performance. Many problems involving optimization are difficult to solve, for example in engineering, science, and business. As such, those problems cannot be resolved deterministically within a plausible amount of time [20]. Currently, the main procedure is to solve those problems using hybrid metaheuristics. The idea of hybridizing metaheuristics dates back to the genesis of metaheuristics themselves [64]. Lately, interest in those hybrids has increased considerably in the field of optimization. This fact can be verified by a large number of effective hybrids proposed to deal with a broad range of problems [20].

Combinations of methods such as metaheuristics, machine learning techniques, mathematical programming, and constraint programming have led to very powerful search algorithms. To provide a big picture of the area, the present section is dedicated to a succinct overview of some hybrid metaheuristics found in the literature. The hybridization type focused on here is the one that combines metaheuristics with other complementary metaheuristics.

Recently, Nguyen et al. [58] introduced a hybrid GP method for dynamic job shop scheduling. The objective of the proposal is to enhance the GP search mechanism by improving the effectiveness of the genetic operators and adding an efficient local search heuristic, named Iterated Local Search (ISL) [73], that can refine the rules evolved by GP. The main motivation is the high probability of generating bad dispatching rules when using only GP's genetic operators. Also, it would be wasteful to evaluate those rules, as they are computationally expensive. Experimental results have confirmed that the hybrid method can evolve rules that are better than those obtained with conventional GP methods.

Following the direction on hybrid GPs, Qi et al. [63] introduced a hybrid GP with PSO, called HGPPSO, which aims to build a novel method to enhance the effectiveness of GP-based algorithms. The proposal performance was tested in the symbolic regression problem of two complex functions. Experimental findings showed that HGPPSO presented superior performance than simple GP in both average convergence generations and convergence times.

Tseng and Liang [76] developed a hybrid strategy that integrates ACO, GA and a Local Search (LS) technique. The work applied the hybrid to solve the Quadratic Assignment Problem. In place of randomly initializing the population, the GA uses a starting population created by ACO. Thus, a better start is provided. The pheromone behaves as a feedback process from the GA to the ACO phase. When the GA phase terminates, the ACO phase resumes control. Next, the ACO uses the updated pheromone to exploit the search space and generate good individuals for the coming execution of the GA phase. Finally, the LS technique is run to enhance the individuals produced by ACO and GA.

Vega-Alvarado et al. [79] introduced a hybrid memetic algorithm that applies the Modified Artificial Bee Colony algorithm (MABC) for global search, and Random Walk variant as a local search method, which is adjusted to deal with design constraints by using an ϵ-constraint scheme. The main objective was to demonstrate the capability of that hybrid algorithm as an alternative technique to solve complex problems, especially on numerical optimization. The results suggested that the proposal can produce fitter solutions for real engineering problems.

2.5 Hybrid Cartesian Genetic Programming Algorithms

Given the relevance of the hybrid metaheuristics that we have seen in the brief review of the previous section, we now undertake an extensive review of the prominent hybrid CGP algorithms recently developed which appeared to be competitive against the existing techniques from literature. First, the motivation behind the development of the hybrid CGP approaches is discussed. Next, those methods and the applications in which they are involved are described in detail.

2.5.1 Motivation

As pointed out in Section 2.4, the use of hybrid formulations has proven to be a promising approach in many domains. Those hybrids have presented better performance and potential competitiveness over traditional methods. In general, hybridization is the joint use of distinct methods to enhance solution performance. As each method is well suited for a specific range of scenarios, the applicability of the hybrid technique is enlarged.

In this direction, the graph-based representation of CGP and its numerous advantages allowed that method to be widely accepted by the scientific community and applied to diverse environments. Nevertheless, there are some scenarios in which the incorporation of auxiliary techniques to CGP proves to be necessary, as the exclusive use of CGP cannot achieve satisfactory performance, and can lead to sub-optimal results. For instance, the inclusion of useful operators from other metaheuristics can enhance CGP performance. Furthermore, replacing CGP's traditional Evolution Strategies scheme by another search mechanism may improve the solution's performance. Finally, the inclusion of another metaheuristic after the execution of CGP for the estimation of numerical coefficients may be a promising alternative. Following that path, recent hybrid approaches combining CGP with other metaheuristics have appeared in the literature and produced improved results. Those studies will be detailed in the following sub-section.

2.5.1.1 CGP combined with ant colony optimization

As previously seen in Section 2.3, CGP uses the evolutionary scheme based on Evolution Strategies (ES) to create offspring solutions [52]. Despite the promising results of ES, researchers have been replacing that approach with other evolutionary mechanisms, in order to evolve the graph representation of CGP and potentially generating better quality programs.

An alternative approach to evolution is Ant Colony Optimization (ACO) [19], a technique for combinatorial problems inspired by ants' cooperativism. ACO has been used in a vast range of combinatorial optimization problems [16], and the significant results found in those problems have led some researchers to also test ACO for the evolution of CGP programs [30, 43].

Hara et al. [30] makes use of the CGP graph representation and the ACO pheromone route selection in order to evolve the topology of graphs. That approach is known as Cartesian Ant Programming (CAP). The ants start from the outputs, and they move probabilistically to the nodes with a number smaller than the node they are currently visiting until the complete graph is generated. The next node to be visited, d, is set probabilistically by

$$p_{z_i,d} = \frac{\tau_{z_i,d}}{\sum_{j \in S_i} \tau_{z_i,j}} \tag{2.8}$$

where z_i is the current position of the ant i, S_i is the set of visitable nodes, $\tau_{z_i,d}$ represents the pheromone level between the actual position z_i and the node d, and $p_{z_i,d}$ is the probability of an ant moving from the actual position z_i to the node d.

The table of pheromones contains the pheromone levels for all nodes, from which the ant can travel to other possible paths. This table is updated according to the iteration-best program's fitness, as

$$\tau_{ij}(t+1) \leftarrow (1-\rho) \cdot \tau_{ij}(t) + \Delta\tau_{ij}^{best} \tag{2.9}$$

where $\tau_{ij}(t)$ is the pheromone level of the path between the node input i and the node output j at iteration t, ρ is the evaporation rate ($0 < \rho < 1$), and $\Delta\tau_{ij}^{best}$ is defined according to the given problem, considering the fitness of the best ant. This update rule is based on an improved ACO version called Max-Min Ant System (MMAS) [74].

As this cycle is repeated, the search around promising graphs is executed. This hybrid CGP algorithm was run to the intertwined spirals classification and symbolic regression applications. The results indicate that CAP exhibited superior performance than CGP when applied to the symbolic regression task, and presented the same performance as CGP during the intertwined spirals classification experiments.

Following that direction, the work presented by Luis and dos Santos [43] also introduced a hybrid CGP algorithm that combines CGP with ACO, called CGP-ACO. In that proposal, the pheromone table has the pheromone levels for every input and function necessary to generate a program node whose values are initialized to τ_d. The inaccessible inputs for a node define their pheromone values as zero. By sampling the table of pheromones, the ants select the appropriate input or function available. The probability p_i that the ant picks a specific input or function i is

$$p_i = \frac{\tau_i}{\sum_{j=1}^{N} \tau_j} \tag{2.10}$$

where τ_i is the pheromone value of an input or function i, and N is the total amount of inputs and functions available.

The pheromone levels are updated according to the n best ants, as given by

$$\tau_{ij}(t+1) \leftarrow \tau_{ij}(t) + \sum_{r=1}^{n-1} \left(w - \frac{w}{n-r} \right) \cdot \Delta\tau_{ij}^{r}(t) + w \cdot \Delta\tau_{best} \tag{2.11}$$

where w is a fixed predefined weight, $\tau_{ij}(t)$ is the pheromone value of input (or function) i at node j at iteration t, $\Delta\tau_{ij}^{r}(t)$ is defined as the rth-best ant's fitness if that ant picks an input (or function) i at node j (otherwise it is defined as zero), and finally $\Delta\tau_{best}$ is the best ant's fitness. This update rule is based on an improved ACO version known as Rank-Based Ant System (AS_{rank}) [6].

The evolvability of the proposed hybrid CGP algorithm was tested under different parameter configurations and compared to CAP [30] in two environments: one with fixed targets and another with moving targets. The experiments have demonstrated that CGP-ACO is highly adaptable to the aforementioned conditions, and both methods are similar regarding the evolved population's average fitness. Also, the proposed hybrid CGP-ACO presented better adaptability in a changing condition by keeping a table of pheromones that causes elevated genotype diversity.

Therefore, both works based on the combination of CGP and ACO are very similar. To sum up, Hara et al. [30] uses the MMAS variant of ACO to evolve the topology of the CGP graphs, whereas Luis and dos Santos [43] uses the AS_{rank} variant of ACO to evolve not only the topology, but also the functions of the CGP graphs.

2.5.1.2 CGP combined with biogeography-based optimization and opposition-based learning

Traditional CGP is executed exclusively with mutation [53], without the sharing of genetic information between individuals, which can lead to a lack of exploitation ability. The incorporation of a proper crossover scheme may increase exploitation and speed up its convergence. Hence, traditional CGP is not an efficacious technique at balancing solutions' exploration and exploitation.

To address that issue, Yazdani and Shanbehzadeh [83] developed a hybrid CGP algorithm named Balanced Cartesian Genetic Programming (BCGP), which combines CGP with Biogeography-Based Optimization (BBO) [69] and Opposition-Based Learning (OBL) [75]. According to Ma and Simon [45], BBO presents a satisfactory exploitability due to the operator of migration. Hence, BCGP incorporates that migration by allowing the genetic sharing between individuals, as in the classical crossover. The migration is implemented as previously described in Sub-section 2.2.2.3. The major benefit of that strategy is the use of a crossover operator without causing any alterations in the CGP original representation, unlike previous works [9, 10] that modified such representation.

Additionally, a novel mutation scheme inspired by OBL [75] was proposed to enhance the CGP exploration capability. OBL is a machine learning concept based on the opposite relationship among entities. Here, BCGP uses the notion of Quasi-Opposition-Based Learning [46], whose main effect is the shrinking of the search space as time increases. In OBL mutation, the value of the k-th gene of the i-th member of the offspring C is denoted by $C_{i,k}$, and it is updated by

$$op_k \leftarrow Min_k^C + Max_k^C - C_{i,k} \tag{2.12}$$

$$M_k \leftarrow \frac{Min_k^C + Max_k^C}{2} \tag{2.13}$$

$$C_{i,k} \leftarrow \begin{cases} \text{round}(M_k + (op_k - M_k) \cdot \text{rand}(0,1)) & \text{if } C_{i,k} < M_k \\ \text{round}(op_k + (M_k - op_k) \cdot \text{rand}(0,1)) & \text{otherwise} \end{cases} \tag{2.14}$$

where Min_k^C and Max_k^C are, respectively, the min and max values allowed for the k-th gene in the offspring C, M_k is the mean point, round(x) is a function that rounds x to the closest integer, and rand$(0,1)$ is a random value uniformly distributed from [0,1). Therefore, OBL mutation utilizes the information of the population to decrease the exploration time [83].

Besides OBL mutation, a simple mutation is also applied, in which the gene value is substituted by a random valid integer. Thus, BCGP uses two different forms of mutation, which tend to enlarge population diversity and improve the exploration capability. The choice of which type of mutation is applied to a given gene is done randomly, where both types have the same probability to be chosen.

The proposed CGP hybrid formulation was applied to symbolic regression experiments and compared to CGP. The tests have shown that BCGP performs better than CGP regarding the convergence speed and accuracy. Therefore, by hybridizing the CGP with BBO and OBL, BCGP provided balanced exploration and exploitation capabilities.

2.5.1.3 CGP combined with differential evolution

Artificial Neural Networks (ANNs) [2] are computational methods inspired by the structures and functions of biological neurons in the brain. They are typically projected as algorithms for handling mathematical, engineering and computational applications. ANNs are composed of a collection of interconnected nodes to perform calculations on input patterns and generate output ones [5]. Training methods are able to modify ANNs' internal structure, usually the weights between the nodes. As such, by adapting their weights following a certain optimization criterion, ANNs can be utilized in a vast range of function approximation tasks, e.g., classification, regression, and feature extraction. A traditional optimization method for ANNs is backpropagation [66], which is a deterministic technique based on gradient descent.

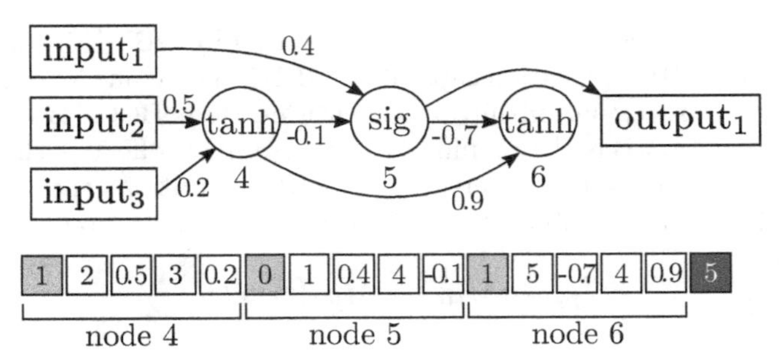

FIGURE 2.3: Illustration of a CGPANN individual with 3 inputs, 3 nodes, and 1 output. Node 6 is inactive. The genotype is presented below the graph. The 3 gray genes represent the transfer functions of the nodes according to a look-up table (0: logistic sigmoid, 1: hyperbolic tangent), the white genes denote the connection genes followed by their corresponding weights, and the final value (dark gray) represents the output connection.

A more recent optimization approach for ANNs is neuroevolution, which is a sub-area of machine learning that executes EAs to train ANNs [77]. In 2010, Khan et al. [39] introduced a neuroevolutionary method called CGPANN in order to use CGP to the training of ANNs. That proposal can optimize the topology and the weights using $(1+\lambda)$-ES. CGPANN extends the traditional CGP by adding connection weight genes for each connection gene, as illustrated in Figure 2.3. It can be noted that the graph representation of CGPANN is suitable to describe ANNs. However, as CGP is designed to handle discrete problems [52], the CGPANN applied in the ANNs' weights can lead to results below the ideal. To circumvent this issue, a better-suited search technique may be used to assist CGP in finding more accurate weights for the ANNs.

Recently, Melo Neto et al. [49] followed that direction and introduced two hybrid CGP algorithms combining CGP and DE for generating ANNs classifiers. In that formulation, CGP evolves the topology while DE optimizes the weights. DE is a stochastic optimizer appropriate to deal with continuous search spaces, so hybrid models should produce more suited weights than the weights generated by the exclusive use of CGP. In that work, the basic DE/rand/1/bin variant is used. The proposed hybrid CGP algorithms were named CGPDE-IN and CGPDE-OUT.

The variant CGPDE-IN alternates the execution between CGP and DE, i.e., DE runs inside the CGP cycle. Before evolution, a special net α is created to store the best individual. During evolution, the child among the λ children that presents the highest fitness is selected, and DE optimizes the connection weight genes of this selected child, maintaining the topology fixed. The starting DE's population (size equals to *Pop*) is composed by the selected child and the other $Pop-1$ nets, that are created utilizing identical topology, but with random weights produced from a uniform distribution. DE returns the best network from the population. If the returned ANN is better than the special solution, then that net is stored as the special individual, replacing the old one. To choose the parent for the coming iteration, the best among the $1 + \lambda$ nets is selected. To create the λ children, λ copies of the parent are created, and mutation is performed exclusively on the topology, which means that the connections are modified, but their weights are kept the same. This cycle repeats until a max number of iterations is reached. Finally, the special net α is chosen as the final solution.

On the other hand, the variant CGPDE-OUT runs DE only after the end of CGP. In summary, CGP is executed and generates a promising topology. After encountering a good topology, DE is executed to evolve the weights that suit that specific topology. Firstly, CGP optimizes the topology by using $(1 + \lambda)$-ES, and along iterations, the best individual is picked. After CGP termination, the weights of the picked individual are optimized by DE. DE produces *Pop* nets with identical topology but distinct weights. When DE is concluded, the method picks the best net in the DE population, which is chosen as the final net.

The proposed hybrid CGP methods were tested on classification problems and compared with backpropagation and CGPANN. The testing findings suggested that those hybrid techniques are statistically superior to the compared techniques.

2.5.1.4 CGP combined with estimation of distribution algorithm

The typical formulation of CGP is composed of integer-valued genes and does not use crossover [77]. Thus, Clegg et al. [10] extended CGP by allowing all genes to be represented as float values in order to perform a floating-point crossover and added an encoding layer, where the genes are represented as a real number from $[0, 1]$. The scheme to decode from floating-point genes to integer-valued genes is performed by using

$$floor(gene_i \cdot function_{total}) \tag{2.15}$$

for the function genes, and

$$floor(gene_i \cdot node_j) \tag{2.16}$$

for the connection genes and output genes, where $gene_i$ is the float number of gene i $(0 \leq i < gene_{total})$, $function_{total}$ is the amount of functions in the look-up table, $gene_{total}$ is the total amount of genes that is present in the genotype, $node_j$ is the node or terminal number $(0 \leq j \leq node_{total})$, $node_{total}$ is the total amount of nodes in the genotype plus the amount of terminals, and $y = floor(x)$ is a function that returns the greatest integer y such that $y \leq x$. Using that approach, crossover was implemented as

$$o_i = (1 - r_i) \cdot p_1 + r_i \cdot p_2, \tag{2.17}$$

where p_1 and p_2 are the genes of the parents, r_i is a random number produced from a uniform distribution, picked for each offspring gene o_i, $0 < r_i < 1$, and $i \in \{1, 2\}$. The mutation scheme alters the gene value to a random real number in the interval [0,1]. That work demonstrated that this CGP extension is more successful in building solutions than the traditional CGPs.

Later on, another work performed by Clegg [9] introduced a hybrid CGP algorithm by combining that floating-point CGP extension with an Estimation of Distribution Algorithm (EDA). A type of EDA called Univariate Marginal Distribution Algorithm (UMDA) [57] was implemented, where its genes present mutually independent probability distributions. To construct new candidate solutions, UMDA uses the components' frequency of a population. That is obtained by assessing the frequency of the components, i.e., the univariate marginal probability [5]. After that, the algorithm makes use of the probabilities in order to induce the probabilistic choice of components in the component-wise building of novel individuals.

In that hybrid proposal, the interval is partitioned into unevenly sized parts that vary regarding some probability distribution. Thus, the functions are not equally likely to be chosen. In the first generation, the method assumes that the genes have a uniform distribution, i.e., the possible values for the genes are identically probable. As the iterations run, those probabilities are increased according to the occurrences of the genes values in promising individuals of the population. As the incremental rate of the probability distribution is a user-defined parameter, several rates have been investigated.

Also, Tournament Selection [3] has been applied to choose the parents. In this scheme, k random members are chosen to participate in the tournament. The best individual from the k individuals is chosen as a parent for the coming iteration. The method repeats this scheme until the desired amount of parents is selected.

The proposed hybrid formulation was compared to the floating-point CGP without EDA on two polynomial regression tasks. Experimental findings have suggested that the incorporation of EDA into the floating-crossover CGP has increased convergence when EDA slowly raises the probabilities as iterations move forward.

2.5.1.5 CGP combined with NSGA-II

A large amount of works have applied CGP to represent and evolve digital circuits [77]. Later on, multi-criteria optimization problems of digital hardware [36, 40] were tackled, and two issues emerged. On the one hand, previous work [37] has shown that global multiobjective genetic algorithms are quite slow when compared to local search methods. On the other hand, local search techniques are not suited to deal with multi-criteria optimization, as they typically work with linear combinations of objectives, and not with the Pareto optimality concept.

To address that issue, Kaufmann et al. [35] introduced a hybrid CGP algorithm, termed hybrid Evolutionary Strategies (hES), which combines the traditional CGP Evolution Strategies search scheme with NSGA-II, a Pareto-based optimizer. That approach can be used for interleaving the run of local and global evolutionary optimizations. The hES uses a $(\mu + \lambda)$-ES on the Pareto-dominant solutions that a multiobjective genetic algorithm has previously acquired. Additionally, it also maintains diversity and avoids the Pareto set deterioration.

The proposed hybrid technique includes two concepts from NSGA-II: (1) the fast non-dominated sorting and (2) the crowding distance, which is used as a metric to preserve diversity. The fast non-dominated sorting is used to pick parents in order to create the offspring. It computes the distinct non-dominated sets for a population of individuals within the objective space. The crowding distance is a metric of how close (in the objectives space) a solution is to its nearest neighbors of the same front [14]. It is utilized to select the offspring candidates to maintain or improve the population diversity of the Pareto set.

The key ideas for hES are described as follows. A local search optimizer is run for each solution of a given set. Only one non-dominated solution from a parent and its offspring move forward to the coming generation. Offspring mutually non-dominated solutions w.r.t. their parent that present different objective vector is ignored. Genetic drift is obtained by ignoring a parent if at least one individual of its offspring presents an identical objective vector. In the end, the parent and its offspring are divided into non-dominated sets using the crowding distance metric.

Every individual in the parent population P_t runs $(1+\lambda)$-ES and appends the recently generated offspring to Q_t. The algorithm then concatenates the parents and offspring using an add-replace function, which copies the parent and adds offspring members that have a single objective vector to the population. If an offspring presents identical objective vector as its parent, it substitutes the parent. By applying the fast non-dominated sorting, the method divides the entire set R_t into non-dominated sets $\mathcal{F}_i$. Next, the method partitions $\mathcal{F}_1$ by the parents into $\mathcal{G} = \{\mathcal{G}_1, \mathcal{G}_2, ...\}$. Here, $\mathcal{G}_i$ is defined as $\mathcal{F}_1 \cap S$ where $S \subseteq P_t \cup Q_t$ contains a parent p and its offspring, where p is a parent of $\mathcal{G}_i$. If a set $\mathcal{G}_i$ does not include the parent p, the least crowded point of $\mathcal{G}_i$ moves forward to the coming iteration. Otherwise, that parent continues to the coming iteration. The remaining individuals of S are removed.

The hybrid CGP algorithm was tested in (i) several MOP benchmark experiments, and (ii) digital hardware evolution, using CGP as the hardware representation model. The hES was compared to NSGA-II and SPEA2 [85], presenting a significantly better performance for most problems.

2.5.1.6 CGP combined with harmony search

Feature processing [28] is one of the necessary steps in predictive analytics. The primary dataset is treated, modified and filtered in order to provide important data characteristics, and reduce the complexity of predictive models. In this direction, feature construction is a class of feature processing techniques that transforms the primary dataset into a feature space of reduced dimensions [28]. In that category, the number of different operator subsets, the particular disposition of the variables within the built characteristic, and the primary characteristics involved in the constructed feature originate a high-dimensional search space. This demands effective algorithms able to evolve programs toward locations of gradually raised optimality [21].

For that purpose, Elola et al. [21] introduced an iterative feature construction scheme based on a hybrid CGP algorithm called Adaptive Cartesian Harmony Search (ACHS), that combines CGP with Harmony Search (HS) [25]. That proposal uses CGP as a solution encoding strategy, suitable at describing characteristics' combinations by using fixed length vectors. In addition, the HS search procedure is applied as the central heuristic of the optimizer. Therefore, that work concentrates on using HS as the heuristic mechanism for evolving CGP programs. ACHS aims at coupling the demonstrated good search capability of the HS with the appropriate constant-length encoding scheme of the CGP. In other words, the role of each CGP individual is to transform a dataset with original features into a dataset with reduced features. The original features serve as the CGP inputs, and the reduced features are the resulting CGP outputs. The evolution of the CGP individuals is performed by HS. In this direction, CGP encoding allows combinations of numerical inputs, keeping the majority of properties expected for a representation to be suitable for HS. Some of these properties are the minimum representation, the fixed length, the redundancy, and the neighboring relation between the values for a specific note driven by its fitness.

The configuration of ACHS depends on a classifier M_θ able to internally predict the target variable, and hence estimating the significance of the input variables. This predicted significance is explored at a wrapper optimizer, which continuously optimizes a population of individuals that describe specific expressions of built variables. The returned dataset from each group of modified characteristics is classified by the learning model. The cross-validated score is utilized to evaluate the quality of the solution, serving as the fitness function that conducts the search procedure [21].

The performance of the hybrid CGP formulation was tested and compared to the current best methods in one toy example and three applications from the literature. Several feature processing criteria and classifiers were used to label data and measure the significance of the input characteristics. The accuracy scores obtained by that hybrid technique were promising, with statistically significant performance gains, which motivates the extensive use of the proposal to supervised learning with datasets constructed by already processed features.

2.5.1.7 CGP combined with particle swarm optimization

In the context of System Biology, the biochemical reactions modeling is a complex procedure. Frequently, there is a knowledge gap regarding the precise molecular mechanisms that occur within the cells. Therefore, several researches that concentrate on the development of Reverse Engineering (RE) techniques have been carried out [1].

In this direction, CGP is an alternative tool to be used for the RE problem, describing/evolving the networks of biochemical reactions by using its graph representation. Besides the evolution of the network structure, an adequate parameter discovery procedure must be associated with the reaction network to capture its behavior. However, the co-evolution of parameters performed by the CGP could, instead, discard a potentially good topology [59]. In fact, Lenser et al. [42] demonstrated that a separation between the networks' evolution and the kinetic parameters' evolution improved the final fitness and also prevented the premature structural convergence. Hence, a better-suited search technique can be used to assist CGP in finding the best set of parameters for the potential networks.

In this sense, Nobile et al. [59] introduced a hybrid CGP method to automatically rebuild the parameterized kinetic reaction network of a specific biological process. That approach combines CGP and Particle Swarm Optimization (PSO) [38], where CGP is applied to find the topology of the network, and PSO estimates the parameters. In that proposal, the PSO individuals represent candidate parameterizations of a reaction-based model previously set through CGP. The PSO is implemented as previously described in Sub-section 2.2.2.8. The basic process of the hybrid CGP-PSO is detailed as follows. The process initializes with the generation of a population of $I = 1 + \lambda$ random CGP programs. Next, the population executes the ES scheme, which can be separated in three steps.

In the first step, each CGP program is transformed into a network of chemical reactions. After, PSO runs for parameter estimation and computes the quality of the parameterized networks. In the second step, the CGP programs are sorted w.r.t. their fitness in order to recognize the best candidate. In the third step, a novel population is created, whose members are the best CGP program of the current population and the $I - 1$ offspring produced by mutation. Furthermore, in order to raise the convergence speed, a copy of the current best network is included in the novel population as the $(I + 1)$-th individual.

That iterative process is repeated. In the end, the best CGP program, including its parameters' configuration, is the outcome of the RE task. The performance of the hybrid CGP approach was tested under a set of biochemical reaction networks for the reverse engineering problem. Experimental findings showed that the developed methodology can rebuild kinetic networks that soundly fit the desired behaviors, obtaining significant performances on the reaction networks investigated.

2.6 Discussion on Hybrid CGP Algorithms

After reviewing the hybrid CGP algorithms, their characteristics, formulations, and applicability, this section is dedicated to an in-depth discussion and critical analysis of those methods.

There are three main axes along which the hybrid formulations can be analyzed: (1) the search schemes, (2) the application context, and (3) the genotype representation of the individuals. All the studies used metaheuristics to evolve the CGP individuals, either by replacing the traditional $(\mu+\lambda)$-ES [9, 21, 30, 43, 83], or by working together with it [35, 49, 59]. The works that replaced ES have applied Ant Colony Optimization [30, 43], Harmony Search [21], UMDA/Tournament [9], or BBO/OBL [83] instead. The works that combined $(\mu+\lambda)$-ES with another metaheuristic have incorporated Differential Evolution [49], NSGA-II [35], or Particle Swarm Optimization [59].

The application context in which the hybrids were used is diverse. Most problems were related to machine learning, i.e., regression [9, 30, 43, 83], classification [30, 49], and feature construction for predictive analytics [21]. The remaining methods were applied to reverse engineering of kinetic reaction networks [59], and digital circuits evolution and MOP benchmarks [35]. Only the approach developed by Kaufmann et al. [35] was dedicated to multiobjective optimization through the use of a Pareto-based technique. Meanwhile, the remaining studies were designed for single-objective optimization only.

Furthermore, two of the eight works have applied the hybridization over a different CGP genotype representation. As such, Melo Neto et al. [49] used the CGP extension for Artificial Neural Networks (CGPANN), which includes the connection weight genes in the genotype. Also, Clegg [9] used the floating-point CGP extension, which encodes the integer-valued genes by real-valued genes. In contrast, the other works have maintained the typical variant of CGP, with the traditional genotype representation.

The above-mentioned analyses are summarized in Table 2.1. In general, the use of different search schemes seems to be a promising strategy for the development of new hybrid CGP algorithms. Also, the use of different CGP genotype representations should be further studied. The diversified applicability of those methods indicates the far-reaching possibilities that can be developed using CGP with other metaheuristics.

2.7 Future Directions of Hybrid CGP Algorithms

Despite the scarcity of papers, research with hybrid CGP algorithms presents a promising direction. There are umpteen open problems and novel

TABLE 2.1: Overview of the hybrid CGP algorithms

Hybrid CGP Algorithm	**Search Scheme**	**Application Context**	**Genotype Representation**
CAP [30]	MMAS	Symbolic Regression and Classification	CGP
CGP-ACO [43]	AS_{rank}	Symbolic Regression	CGP
BCGP [83]	BBO + OBL	Symbolic Regression	CGP
CGPDE [49]	ES + DE	Classification	CGPANN
CGP-EDA [9]	UMDA + Tournament	Polynomial Regression	Floating-CGP
hES [35]	ES + NSGA-II	Hardware Evolution and MOP Benchmarks	CGP
ACHS [21]	HS	Feature Construction	CGP
CGP-PSO [59]	ES + PSO	Reverse Engineering of Reaction Networks	CGP

areas of application that are potentially appropriate for using this approach. The present section outlines important future research directions in the area of hybrid CGP algorithms.

Firstly, the use of other evolutionary schemes, either by replacing the traditional ES or by working together with it, seems to be an interesting approach for further investigation. Many metaheuristics are available in the literature and can potentially produce significant results when combined with CGP. For instance, Motta et al. [56] recently proposed a hybrid approach of Grammar-based GP (GGP) with ES to enhance the performance of symbolic regression applications. GGP optimizes the models' structure and ES searches for the coefficients. Alternatively, that hybrid can be modified by replacing GGP by CGP, which may also lead to good results.

Hybrid CGP formulations using deterministic methods to aid the optimization process can also be developed. For instance, the CGP variant for ANNs (CGPANN) can optimize the connection weights by applying the backpropagation algorithm or other training method related to Machine Learning (ML). Other neuroevolutionary algorithms, such as NEAT [7], have applied that principle by incorporating backpropagation and showed good performance. Regarding the use of hybrid CGP in the ML environment, deep learning algorithms [27] have received great attention due to their performance breakthroughs. In that direction, those hybrid models can be devised to aid both the structural representation of deep networks as well as their training processes.

Also, the effectiveness of CGP depends on the selected mutation strategy [77]. However, conventional CGP has a limited set of mutation strategies.

Thus, the inclusion of additional search techniques for designing new mutation operators is relevant. In addition, many metaheuristics based on genetic sharing of individuals can be combined with CGP to allow for an increased exploitability of the search space, similar to the work performed by Yazdani and Shanbehzadeh [83]. As crossover operations have gained little attention in CGP [52], future hybrid approaches focused on the use of recombination operators from other metaheuristics deserve further investigation.

As an example, a recent crossover operator for CGP was proposed by Silva and Bernardino [68] for designing combinational logic circuits, in which the standard ES was modified by gathering the graphs of the best parent's outputs and its offspring, so it can build a potentially fitter candidate member. Although that crossover operator has originated from ES itself and not from another metaheuristic, that work illustrates the many modifications that can be further explored to create new variation operators, either by modifying ES or by hybridizing it with other techniques.

Furthermore, inspired by Clegg [9] and Melo Neto et al. [49], it was verified that the combined use of different CGP genotypes with other search methods has presented significant results and should be more deeply studied. Given the large number of available CGP extensions, such as Self-Modifying CGP [32] or Modular CGP [81], progress can also be achieved if those extensions incorporate other metaheuristics, as they can potentially enhance CGP's performance.

Moreover, some CGP parameters could be fixed by the user before the search, for example, the number of nodes, arity, and levels-back. As an alternative approach, those parameters can be fine-tuned by using other search techniques as meta-optimizers, in order to adapt CGP to a given problem domain. Other CGP parameters may vary during the search, e.g., the mutation rate. In such cases, one can think of adaptive processes to automatically define the values of these other parameters by using feedback from the search process.

Inspired by Kaufmann et al. [35], further investigation on the use of hybrid CGP to solve Multiobjective Problems (MOPs) is encouraged. In general, EAs are well suited for solving MOPs, as they maintain a population of individuals [11] which approximate the Pareto optimal set in a single algorithm execution. In this direction, the combination of CGP with multiobjective EAs (MOEAs), e.g., SPEA2 [85], may lead to meaningful solutions when handling such problems.

Finally, the development of hybrid CGPs to deal with many objectives should also be considered as a promising research direction. Those problems typically deal with more than three objectives and, as Das and Suganthan [13] pointed out, the performance of conventional MOEAs that apply Pareto domination as a ranking metric usually decrease sharply, as the amount of objectives grows. Thus, the hybridization of CGP with other search methods to handle many-objective tasks remains open as a challenging area of research.

2.8 Concluding Remarks

The growing complexity of real-world optimization tasks demands research for fast, robust, and effective optimizers. In this direction, hybrid Cartesian Genetic Programming (CGP) algorithms emerged as efficient and promising schemes for global and local optimization.

This chapter sought to provide an overall picture of the current research on hybrid CGP algorithms. By introducing the metaheuristics, it discussed the essential concepts of the main algorithms used in the hybrid formulations. Next, it gave an extensive review of the CGP method including the historical context, encoding scheme, evolution, parameters, advantages, and drawbacks. Also, a brief overview of other hybrid algorithms found in the literature gives a big picture of the area. A review on the latest hybrid CGP algorithms was performed, with a detailed description of the techniques and results found when applied to their application domain.

Then, it provided critical analysis and an in-depth discussion on those hybrid CGP approaches, highlighting the similarities between those works. That analysis was performed along three main axes: (1) the search engines, (2) the application context, and (3) the genotype representation. Regarding the search engines, we notice a diversified set of metaheuristics used to optimize the CGP models. Regarding the applications, machine learning was the field that has received the most attention from those algorithms. Also, the traditional CGP genotype representation was the most often adopted among the hybrids.

Finally, this review unfolded a number of future research directions for hybrid CGPs such as: the use of other stochastic or deterministic optimizers as CGP search engines, the opportunity of studying novel CGP variation operators inspired by other metaheuristics, the use of different genotype representations, and the application of hybrid CGPs to multiobjective problems. The structural and functional flexibility of CGP, combined with different search schemes, has led to significant results and should be further investigated and applied in different domains. The contents of the chapter suggest that hybrid CGP algorithms promise to be a vibrant area of multidisciplinary research in the near future.

Acknowledgment

This study was financed in part by the Coordenação de Aperfeiçoamento de Pessoal de Nível Superior - Brasil (CAPES) - Finance Code 001. The authors also thank the financial support provided by FAPEMIG (grants APQ-03414-15 and APQ-00337-18) and CNPq (grants 312337/2017-5 and 312682/2018-2).

Bibliography

[1] J. Ashworth, E. Wurtmann, and N. Baliga. Reverse engineering systems models of regulation: discovery, prediction and mechanisms. *Current Opinion in Biotechnology*, 23(4):598–603, 2012.

[2] C. Bishop et al. *Neural Networks for Pattern Recognition.* Oxford University Press, 1995.

[3] T. Blickle and L. Thiele. A mathematical analysis of tournament selection. In *Proceedings of the Sixth International Conference on Genetic Algorithms*, pages 9–16. Morgan Kaufmann, 1995.

[4] S. Brooks. A discussion of random methods for seeking maxima. *Operations Research*, 6(2):244–251, 1958.

[5] J. Brownlee. *Clever Algorithms: Nature-Inspired Programming Recipes.* Lulu.com, 1st edition, 2011.

[6] B. Bullnheimer, R. Hartl, and C. Strauss. A new rank based version of the ant system - A computational study. *Central European Journal of Operations Research*, 7:25–38, 1999.

[7] L. Chen and D. Alahakoon. Neuroevolution of augmenting topologies with learning for data classification. In *2006 International Conference on Information and Automation*, pages 367–371, Dec 2006.

[8] T. Clarke. *Cartesian Genetic Programming for Control Engineering*, pages 157–173. Springer International Publishing, 2018.

[9] J. Clegg. Combining cartesian genetic programming with an estimation of distribution algorithm. In *Proc. of the 10th Annual Conf. on Genetic and Evolutionary Computation*, pages 1333–1334, New York, NY, USA, 2008. ACM.

[10] J. Clegg, J. Walker, and J. Miller. A new crossover technique for cartesian genetic programming. In *Proc. of the 9th Annual Conf. on Genetic and Evolutionary Computation*, pages 1580–1587, New York, NY, USA, 2007. ACM.

[11] C. Coello, G. Lamont, and D. Veldhuizen. *Evolutionary Algorithms for Solving Multi-Objective Problems (Genetic and Evolutionary Computation).* Springer-Verlag, Berlin, Heidelberg, 2006.

[12] N. Cramer. A representation for the adaptive generation of simple sequential programs. In *Proceedings of the 1st International Conference on Genetic Algorithms*, pages 183–187. L. Erlbaum Associates Inc., 1985.

[13] S. Das and P. Suganthan. Differential evolution: A survey of the state-of-the-art. *IEEE Transactions on Evolutionary Computation*, 15(1):4–31, Feb 2011.

[14] K. Deb, S. Agrawal, A. Pratap, and T. Meyarivan. A fast elitist non-dominated sorting genetic algorithm for multi-objective optimization: NSGA-II. In M. Schoenauer, K. Deb, G. Rudolph, X. Yao, E. Lutton, J. Merelo, and H. Schwefel, editors, *Parallel Problem Solving from Nature PPSN VI*, pages 849–858. Springer Berlin Heidelberg, 2000.

[15] M Dorigo. The ant system: optimization by a colony of cooperating agents. *IEEE Transactions on Systems, Man, and Cybernetics-Part B*, 26(1):1–13, 1996.

[16] M. Dorigo and C. Blum. Ant colony optimization theory: A survey. *Theoretical Computer Science*, 344(2):243–278, 2005.

[17] M. Dorigo and L. Gambardella. Ant colony system: a cooperative learning approach to the traveling salesman problem. *IEEE Transactions on Evolutionary Computation*, 1(1):53–66, 1997.

[18] M. Dorigo and T. Stützle. *The Ant Colony Optimization Metaheuristic: Algorithms, Applications, and Advances*, pages 250–285. Springer US, Boston, MA, 2003.

[19] M. Dorigo and T. Stützle. *Ant Colony Optimization.* Bradford Company, Scituate, MA, USA, 2004.

[20] E. Talbi (ed.). *Hybrid Metaheuristics.* Studies in Computational Intelligence. Springer Berlin Heidelberg, 1st edition, 2013.

[21] A. Elola, J. Del Ser, M. Bilbao, C. Perfecto, E. Alexandre, and S. Salcedo-Sanz. Hybridizing cartesian genetic programming and harmony search for adaptive feature construction in supervised learning problems. *Applied Soft Computing*, 52:760–770, 2017.

[22] S. Forrest and M. Mitchell. Relative building-block fitness and the building-block hypothesis. In *Foundations of Genetic Algorithms*, volume 2, pages 109–126. Elsevier, 1993.

[23] R. Friedberg. A learning machine: Part i. *IBM Journal of Research and Development*, 2:2–13, 1958.

[24] Z. Geem. *Music-Inspired Harmony Search Algorithm: Theory and Applications*, volume 191. Springer, 2009.

[25] Z. Geem, J. Kim, and G. Loganathan. A new heuristic optimization algorithm: Harmony search. *Simulation*, 76(2):60–68, 2001.

[26] F. Glover and M. Laguna. Tabu search. In D. Du and P. Pardalos, editors, *Handbook of Combinatorial Optimization: Volume1–3*, pages 2093–2229. Springer US, Boston, MA, 1999.

[27] I. Goodfellow, Y. Bengio, and A. Courville. *Deep Learning.* MIT Press, 2016.

[28] J. Han, M. Kamber, and J. Pei. Data preprocessing. In J. Han, M. Kamber, and J. Pei, editors, *Data Mining*, pages 83–124. Morgan Kaufmann, Boston, third edition, 2012.

[29] A. Hara, H. Konishi, J. Kushida, and T. Takahama. Efficiency improvement of imitation operator in multi-agent control model based on cartesian genetic programming. In *2016 IEEE International Workshop on Computational Intelligence and Applications*, pages 69–74, Nov 2016.

[30] A. Hara, M. Watanabe, and T. Takahama. Cartesian ant programming. In *2011 IEEE International Conference on Systems, Man, and Cybernetics*, pages 3161–3166, Oct 2011.

[31] S. Harding, J. Leitner, and J. Schmidhuber. *Cartesian Genetic Programming for Image Processing*, pages 31–44. Springer New York, 2013.

[32] S. Harding, Julian Miller, and W. Banzhaf. Self-modifying cartesian genetic programming. In J. Miller, editor, *Cartesian Genetic Programming*, pages 101–124. Springer Berlin Heidelberg, 2011.

[33] John H. Holland. *Adaptation in Natural and Artificial Systems: An Introductory Analysis with Applications to Biology, Control and Artificial Intelligence.* MIT Press, Cambridge, MA, USA, 1992.

[34] A. Hopgood, editor. *Intelligent Systems for Engineers and Scientists (2nd Ed.).* CRC Press, Inc., Boca Raton, FL, USA, 2001.

[35] P. Kaufmann, T. Knieper, and M. Platzner. A novel hybrid evolutionary strategy and its periodization with multi-objective genetic optimizers. In *IEEE Congress on Evolutionary Computation*, pages 1–8, July 2010.

[36] P. Kaufmann and M. Platzner. Toward self-adaptive embedded systems: Multi-objective hardware evolution. In P. Lukowicz, L. Thiele, and G. Tröster, editors, *Architecture of Computing Systems - ARCS 2007*, pages 199–208. Springer Berlin Heidelberg, 2007.

[37] P. Kaufmann, C. Plessl, and M. Platzner. Evocaches: Application-specific adaptation of cache mappings. In *2009 NASA/ESA Conference on Adaptive Hardware and Systems*, pages 11–18, July 2009.

[38] J. Kennedy and R. Eberhart. Particle swarm optimization. In *Proceedings of ICNN'95 - International Conference on Neural Networks*, volume 4, pages 1942–1948, Nov 1995.

[39] M. Khan, G. Khan, and J. Miller. Evolution of neural networks using cartesian genetic programming. In *IEEE Congress on Evolutionary Computation*, pages 1–8, July 2010.

[40] T. Knieper, B. Defo, P. Kaufmann, and M. Platzner. On robust evolution of digital hardware. In M. Hinchey, A. Pagnoni, F. Rammig, and H. Schmeck, editors, *Biologically-Inspired Collaborative Computing*, pages 213–222, Boston, MA, 2008. Springer US.

[41] J. Koza. *Genetic Programming: On the Programming of Computers by Means of Natural Selection.* MIT Press, Cambridge, MA, USA, 1992.

[42] T. Lenser, T. Hinze, B. Ibrahim, and P. Dittrich. Towards evolutionary network reconstruction tools for systems biology. In E. Marchiori, J. Moore, and J. Rajapakse, editors, *Evolutionary Computation, Machine Learning and Data Mining in Bioinformatics*, pages 132–142. Springer Berlin Heidelberg, 2007.

[43] S. Luis and M. dos Santos. On the evolvability of a hybrid ant colony-cartesian genetic programming methodology. In K. Krawiec, A. Moraglio, T. Hu, A. Etaner-Uyar, and B. Hu, editors, *Genetic Programming*, pages 109–120. Springer Berlin Heidelberg, 2013.

[44] S. Luke. *Essentials of Metaheuristics.* Lulu, second edition, 2013.

[45] H. Ma and D. Simon. Blended biogeography-based optimization for constrained optimization. *Engineering Applications of Artificial Intelligence*, 24(3):517–525, 2011.

[46] S. Mahdavi, S. Rahnamayan, and K. Deb. Opposition based learning: A literature review. *Swarm and Evolutionary Computation*, 39:1–23, 2018.

[47] K. Man, K. Tang, and S. Kwong. Genetic algorithms: concepts and applications [in engineering design]. *IEEE Transactions on Industrial Electronics*, 43(5):519–534, Oct 1996.

[48] D. Manjarres, I. Landa-Torres, S. Gil-Lopez, J. Del Ser, M.N. Bilbao, S. Salcedo-Sanz, and Z.W. Geem. A survey on applications of the harmony search algorithm. *Engineering Applications of Artificial Intelligence*, 26(8):1818–1831, 2013.

[49] J. Melo Neto, H. Bernardino, and H.J.C. Barbosa. Hybridization of cartesian genetic programming and differential evolution for generating classifiers based on neural networks. In *2018 IEEE Congress on Evolutionary Computation*, pages 614–621, July 2018.

[50] Z. Michalewicz, C. Janikow, and J. Krawczyk. A modified genetic algorithm for optimal control problems. *Computers & Mathematics with Applications*, 23(12):83–94, 1992.

[51] J. Miller. An empirical study of the efficiency of learning boolean functions using a cartesian genetic programming approach. In *Proc. of the 1st Annual Conf. on Genetic and Evolutionary Computation*, pages 1135–1142, San Francisco, CA, USA, 1999. Morgan Kaufmann.

[52] J. Miller. *Cartesian Genetic Programming.* Springer-Verlag Berlin Heidelberg, 2011.

[53] J. Miller, D. Job, and V. Vassilev. Principles in the evolutionary design of digital circuits—Part I. *Genetic Programming and Evolvable Machines*, 1(1):7–35, Apr 2000.

[54] J. Miller and P. Thomson. Cartesian genetic programming. In R. Poli, W. Banzhaf, W. Langdon, J. Miller, P. Nordin, and T. Fogarty, editors, *Genetic Programming*, pages 121–132. Springer Berlin Heidelberg, 2000.

[55] J.F. Miller, P. Thomson, and T. Fogarty. Designing electronic circuits using evolutionary algorithms. Arithmetic circuits: A case study. In D. Quagliarella, J. Periaux, C. Poloni, and G. Winter, editors, *Genetic Algorithms and Evolution Strategies in Engineering and Computer Science: Recent Advances and Industrial Applications*, pages 105–128. John Wiley & Sons, 1998.

[56] F. Motta, J. Freitas, F. Souza, H. Bernardino, I. Oliveira, and H.J.C Barbosa. A hybrid grammar-based genetic programming for symbolic regression problems. In *2018 IEEE Congress on Evolutionary Computation (CEC)*, pages 1–8, July 2018.

[57] H. Mühlenbein. The equation for response to selection and its use for prediction. *Evolutionary Computation*, 5(3):303–346, 1997.

[58] S. Nguyen, Y. Mei, B. Xue, and M. Zhang. A hybrid genetic programming algorithm for automated design of dispatching rules. *Evolutionary Computation*, pages 1–31, June 2018.

[59] M. Nobile, D. Besozzi, P. Cazzaniga, D. Pescini, and G. Mauri. Reverse engineering of kinetic reaction networks by means of cartesian genetic programming and particle swarm optimization. In *2013 IEEE Congress on Evolutionary Computation*, pages 1594–1601, June 2013.

[60] M. Pelikan. *Probabilistic Model-Building Genetic Algorithms*, pages 13–30. Springer, Berlin, Heidelberg, 2005.

[61] M. Pelikan, D. Goldberg, and E. Cantu-Paz. Linkage problem, distribution estimation, and bayesian networks. *Evolutionary Computation*, 8(3):311–340, 2000.

[62] R. Poli, W. Langdon, and N. McPhee. *A Field Guide to Genetic Programming.* Lulu Enterprises, UK Ltd, 2008.

[63] F. Qi, Y. Ma, X. Liu, and G. Ji. A hybrid genetic programming with particle swarm optimization. In Y. Tan, Y. Shi, and H. Mo, editors, *Advances in Swarm Intelligence*, pages 11–18. Springer Berlin Heidelberg, 2013.

[64] G. Raidl. A unified view on hybrid metaheuristics. In F. Almeida, M. Aguilera, C. Blum, J. Marcos M. Vega, M. Pérez, A. Roli, and M. Sampels, editors, *Hybrid Metaheuristics*, pages 1–12. Springer Berlin Heidelberg, 2006.

[65] I. Rechenberg. *Evolutionsstrategie: Optimierung technischer Systeme nach Prinzipien der biologischen Evolution.* PhD thesis, TU Berlin, 1971.

[66] D. Rumelhart, G. Hinton, and R. Williams. Learning representations by back-propagating errors. *Nature*, 323(6088):533, 1986.

[67] A. Adamatzky and S. Stepney (eds.). *Inspired by Nature: Essays Presented to Julian F. Miller on the Occasion of his 60th Birthday.* Emergence, Complexity and Computation 28. Springer International Publishing, 1st edition, 2018.

[68] J. Silva and H. Bernardino. Cartesian genetic programming with crossover for designing combinational logic circuits. *7th Brazilian Conference on Intelligent Systems*, pages 1–6, 2018.

[69] D. Simon. Biogeography-based optimization. *IEEE Transactions on Evolutionary Computation*, 12(6):702–713, Dec 2008.

[70] S. Smith. *A Learning System Based on Genetic Adaptive Algorithms.* PhD thesis, University of Pittsburgh, Pittsburgh, PA, USA, 1980.

[71] N. Srinivas and K. Deb. Multiobjective optimization using nondominated sorting in genetic algorithms. *Evolutionary Computation*, 2(3):221–248, 1994.

[72] R. Storn and K. Price. Differential evolution–a simple and efficient adaptive scheme for global optimization over continuous spaces. Technical Report TR-95-012, ICSI, USA, 1995.

[73] T. Stützle. *Local Search Algorithms for Combinatorial Problems: Analysis, Improvements, and New Applications.* PhD thesis, Darmstadt University of Technology, 1998.

[74] T. Stützle and H. Hoos. Max–min ant system. *Future Generation Computer Systems*, 16(8):889–914, 2000.

[75] H. Tizhoosh. Opposition-based learning: A new scheme for machine intelligence. In *Intl. Conference on Computational Intelligence for Modelling, Control and Automation and Intl. Conference on Intelligent Agents, Web Technologies and Internet Commerce*, volume 1, pages 695–701, Nov 2005.

[76] L. Tseng and S. Liang. A hybrid metaheuristic for the quadratic assignment problem. *Computational Optimization and Applications*, 34(1):85–113, May 2006.

[77] A. Turner. *Evolving Artificial Neural Networks using Cartesian Genetic Programming*. PhD thesis, University of York, 2015.

[78] F. Ullah. American sign language recognition system for hearing impaired people using cartesian genetic programming. In *The 5th Intl. Conference on Automation, Robotics and Applications*, pages 96–99, Dec 2011.

[79] E. Vega-Alvarado, E. Portilla-Flores, M. Calva-Yáñez, G. Sepúlveda-Cervantes, J. Aponte-Rodríguez, E. Santiago-Valentín, and J. Rueda-Meléndez. Hybrid metaheuristic for designing an end effector as a constrained optimization problem. *IEEE Access*, 5:6002–6014, 2017.

[80] P. Vikhar. Evolutionary algorithms: A critical review and its future prospects. In *2016 Intl. Conf. on Global Trends in Signal Processing, Information Computing and Communication*, pages 261–265, Dec 2016.

[81] J. Walker and J. Miller. The automatic acquisition, evolution and reuse of modules in cartesian genetic programming. *IEEE Transactions on Evolutionary Computation*, 12(4):397–417, Aug 2008.

[82] D. Weyland. A critical analysis of the harmony search algorithm—how not to solve sudoku. *Operations Research Perspectives*, 2:97–105, 2015.

[83] S. Yazdani and J. Shanbehzadeh. Balanced cartesian genetic programming via migration and opposition-based learning: application to symbolic regression. *Genetic Programming and Evolvable Machines*, 16(2):133–150, Jun 2015.

[84] Y. Zhang, S. Wang, and G. Ji. A comprehensive survey on particle swarm optimization algorithm and its applications. *Mathematical Problems in Engineering*, 2015, 2015.

[85] E. Zitzler, M. Laumanns, and L. Thiele. SPEA2: Improving the strength Pareto evolutionary algorithm. *TIK-report*, 103, 2001.

Chapter 3

Tuberculosis Detection from Conventional Sputum Smear Microscopic Images Using Machine Learning Techniques

Rani Oomman Panicker
(a) Department of Computer Applications, Cochin University of Science and Technology, Kochi, India
(b) Department of Information Technology, College of Engineering, Trikaripur, Kasaragod, India

Biju Soman
AMCHSS, Sree Chitra Tirunal Institute for Medical Sciences and Technology, Trivandrum, India

M.K. Sabu
Department of Computer Applications, Cochin University of Science and Technology, Kochi, India

3.1 Introduction 63
3.2 Sputum Smear Microscopic Images 65
3.2.1 Disadvantages of Conventional Methods 66
3.3 Machine Learning Techniques for TB Detection 67
3.3.1 Study Design 68
3.3.2 Literature Review 68
3.4 Discussion 76
3.5 Conclusions and Future Scope 77
Bibliography 78

3.1 Introduction

Tuberculosis (TB) is a serious contagious disease that spreads from one person to another through air. Every second, a new person in the world gets infected with TB disease [25]. A type of slow growing bacterium, Mycobacterium tuberculosis (MTB) causes TB and these bacilli are rod shaped, with varying lengths from 1 μm to 10 μm [13]. It mainly affects the lungs, but it can also attack other parts of the human body such as bone marrow, kidney, spinal cord, brain, etc. [13]. The former type of TB disease is called pulmonary TB and the latter type is called extra pulmonary TB [14]. The symptoms (fever, cough, weight loss, night sweats, etc.) of TB may not be serious for several months, and an infected person can spread it to 10 to 15 other people yearly through direct communication. As per the World Health Organization (WHO) 2018 report, 10 million people are affected with TB, among them 1.3 million died in 2017 [7]. Also, two-thirds of global TB load is concentrated in eight countries including India (27%), China (9%), Indonesia (8%), etc. [14]. So, WHO treated the TB disease as a global emergency [14, 25] and also recommended the urgent need of adoption of new technology for reducing TB deaths and new TB incidence [7].

TB is mainly classified into two types: latent TB and active TB. In latent TB, the bacilli are seen in an inactive or idle state and become active in the future, whereas active TB is a serious case and can infect others. TB is detected by using conventional (bright field) and fluorescent microscope. The main difference between the two is its staining procedure; in fluorescent microscope auramine-o staining procedure is used, whereas in conventional microscope Ziehl-Neelsen (ZN) staining is used [13]. Compared to conventional microscope, fluorescent microscope is costly and needs well-trained technicians for its operation and management; so the majority of developing countries like India and Africa use bright field microscope for TB detection [13].

The majority of active TB cases can be cured when the right medication and treatments are given; otherwise it may turn fatal. Different diagnostic methods such as chest X-ray, microscopic sputum smear examination, culture test, tuberculin skin test, GeneXpert, interferon-γ release assay (IGRA), etc. are available for TB detection [13]. But sputum smear microscopic examination with bright field or conventional microscope is the commonly followed method for the detection of pulmonary TB, as this is comparatively cost effective and easy to use [13]. Unfortunately, manual detection of TB is time consuming, error prone and labor-intensive. A clinician will take a long time (15 min to 3 hours) for assessing at least 100 oil immersion fields (as per the WHO recommendation) in each slide; therefore, one can screen only a limited number of slides in a day [13]. According to Veropoulos et al. [10] the majority of active TB cases may be missed during the manual screening method, but the accuracy in bacilli detection can be improved with the help of automated TB detection systems, as these systems can identify active TB cases very quickly by screening all the hundred fields in each slide [10]. Many researchers developed semi-automated algorithms for detecting TB [4, 10, 14, 17, 20] but only very few researchers developed fully automatic TB detection systems (e.g., TBDX system) [8, 26]. These automated systems use various image-processing techniques for detecting TB bacilli. Also, a majority of conventional studies [6, 11, 24] followed color-based segmentation for extracting the TB bacilli, since bacilli appear in red after the staining procedure. Remaining studies used either different segmentation methods or feature extraction-based methods for extracting TB bacilli from sputum smear microscopic images. But these conventional methods became unsuccessful in identifying occluded or irregular shaped or touching TB bacilli. Also, these methods showed poor detection accuracy; and they do not work well with the sputum smear images with large varied backgrounds. Moreover, the severity of TB disease always depends upon the count of single and touching bacilli, so the inclusion of all types of bacilli in TB detection is very important.

As mentioned earlier, due to the complex nature of sputum smear microscopic images, it is very difficult to extract irregular shaped objects. Therefore, an accurate detection of all types of bacilli requires a method called "learning from examples", which is actually pointing towards machine learning. Presently a machine learning technique plays a vital role in the field of medicine such as computer-aided diagnosis, robotic surgery, drug discovery etc. It is claimed that different types of TB bacilli (single bacilli and touching bacilli) in sputum smear microscopic images can be easily and accurately recognized with the help of machine learning techniques [14]. Many recent machine learning studies [9, 14, 15] also showed better performance in detecting all types of TB bacilli from sputum smear microscopic images compared to other conventional methods.

The aim of this chapter is to cover a wide spectrum of machine learning techniques that are used for recognizing TB bacilli from conventional sputum smear microscopic images. This chapter is organized as follows. Section 3.2 discusses sputum smear microscopic images. Major disadvantages of conventional methods are mentioned in Section 3.2.1 and Section 3.3 discusses the major works that are based on machine learning techniques. Major discussions are mentioned in Section 3.4. Finally, the conclusions and future research scope are drawn in Section 3.5.

3.2 Sputum Smear Microscopic Images

Sputum smear microscopic examination is a simple and low-cost method. This method is one of the most efficient and commonly used approaches for the diagnosis and management of pulmonary TB in many parts of the world. Nowadays, active research is going on for developing various automated diagnosing systems for the treatment of diseases like malaria, cancer, etc. and among them automated diagnosing systems for TB play a major role in effective detection of TB. A clinician collects a patient's early morning sputum and is stained by using either Ziehl-Neelsen (ZN) staining in the case of conventional microscopy or following Auramine-Rhodamine staining in the case of fluorescent microscopy [13]. In ZN staining, the bacillus appears as red against the blue background; otherwise, bacillus is seen as yellow against a black background. In automated setup, after staining, the slide is put under the microscopic eyepiece and captures the image using a digital imager or camera, which is connected to the microscope; and these images can then be saved on the computer for further processing. It is then analyzed using image processing or machine learning techniques to classify the image as TB positive or TB negative.

Sputum smear microscopic images may contain 'no bacilli', 'single or few bacilli', 'touching or overlapping bacilli', and 'over stained with artefacts and bacilli' [12]. Figure 3.1 shows different types of sputum smear microscopic images. Based on the background, these images are divided into two groups, low density and high density background images (LDBI and HDBI). In the former case, the background is seen as light blue or white, whereas in the latter case the background is seen as deep blue; it is due to the light and heavy presence of methylene blue solution [5]. Also, in low density images, the total count of bacilli is much less compared to high density images. If the smear-slide has any variation in staining (i.e, over-staining or under-staining), it will affect the image quality and this will lead to poor detection of TB disease. So, there is a need for some preprocessing techniques for enhancing the quality of such images. As the quality of the sputum smear image increases, the detection accuracy of TB using an automated method also increases.

3.2.1 Disadvantages of Conventional Methods

The first automatic TB detection method based on conventional approaches was proposed by Costa et al. [6] in 2008. Their method utilized color-based segmentation in detecting TB bacilli. Similarly, majority of researchers in the field of automatic TB detection followed color-based segmentation since color is the most crucial factor for determining the bacilli identification process. After segmentation, most of the authors followed some feature extraction techniques and morphological operations for the accurate detection of TB bacilli. But the accuracy of these conventional methods was not that impressive, i.e., accuracy of majority of methods is less than 90% [13], which might not be clinically acceptable. Moreover, all the conventional methods failed in detecting touching, overlapping or occluded bacilli, since these bacilli have irregular shapes and the majority of researchers classified it as non-bacilli. This will adversely affect the performance of those methods. Also, the majority of authors followed their own TB databases, since no common public database existed up to the year 2013. So comparing the methods based on the results reported in those papers is unfair.

The first public database was released by Costa et al. [5] in 2014; and it includes an *auto-focus database* with 1200 images and *segmentation and classification database* with 120 images [5]. All images in the database have 2816 x 2112 pixel resolution and the ground truth of both high density and low density images are also marked by an expert technician. The second public free database (ZNSM-iDB) is made available by Shah et al. [12] in 2017 and it

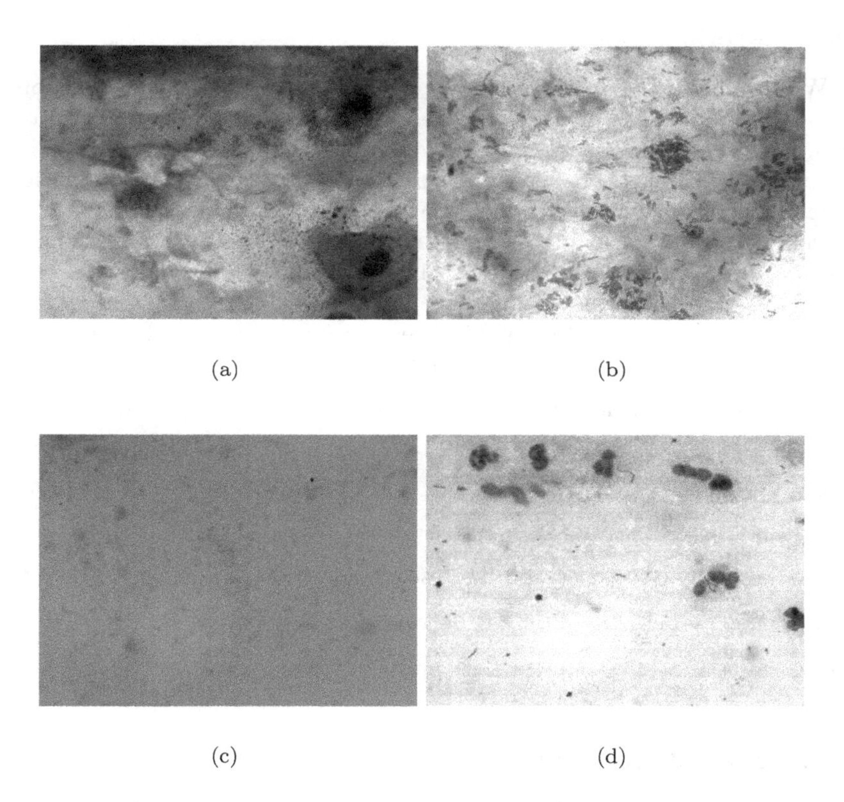

(a) (b)

(c) (d)

FIGURE 3.1: (a) Over-stained with bacilli and artifacts, (b) overlapping or touching or occluded bacilli, (c) no bacilli, (d) single or a few bacilli [Source: [12]]

contains auto-stitching database, auto-focusing database, and segmentation and classification database. These free datasets motivated researchers in developing tools and algorithms for automatic TB detection using image processing and machine learning techniques. Treatment of TB disease always depends on the infection level [11], so accuracy of the detection is very important. Therefore, developments of effective methods are required where machine learning techniques play an important role.

3.3 Machine Learning Techniques for TB Detection

This section discusses the study design and provides a short review of various machine learning methods for TB bacilli detection.

3.3.1 Study Design

We searched various database, like MEDLINE, Google Scholar, Pubmed, etc. and downloaded all the published literature related to automatic TB detection. The keywords used for the search strategy are "automatic TB detection", " automatic TB detection using image processing", "automatic TB detection using machine learning techniques", "Tuberculosis (TB) OR conventional sputum smear microscopic images". The study design flowchart is shown in Figure 3.2 and it contains different inclusion and exclusion criteria of various studies. In this chapter, we included 16 relevant studies which are based on automatic detection of TB disease using conventional sputum smear microscopic images. The exclusion criteria were:

1. Automatic TB detection studies that are based on conventional methods.
2. Automatic TB detection studies that are based on tissue slide image.
3. Studies that describes the TB database only.
4. Studies that are based on fluorescent microscopic images.
5. Review papers based on TB detection.
6. Studies that do not meet the quality.

3.3.2 Literature Review

As mentioned earlier, the first automatic TB detection work based on traditional methods was published by Costa et al. [6] in 2008, and they used bright field sputum smear microscopic images. From 2008 onwards, many researchers put their effort in detecting TB disease automatically; they either followed traditional image processing methods or machine learning techniques. In this section, we present an overview of different automatic methods based on machine learning techniques published between 2008 and 2018.

(a) Bayesian-based methods Sadaphal et al. [17] were the first to suggest a machine learning technique for detecting TB bacilli. The authors adopted a Bayesian segmentation method for predicting the probability of TB bacilli pixel and then classified the bacilli as TB and non-TB based on their shape features such as area, eccentricity and axis ratio. Sadaphal et al. [17] faced so many challenges such as detection of bacilli clusters, over-staining and under-staining of images, etc. i.e., their method failed in detecting T or Y shaped touching bacilli. In 2012, Rulaningtyas et al. [20] used a Naïve Bayes classifier for detecting TB bacilli from sputum smear images. The authors cropped the TB bacilli from sputum smear microscopic images and trained using naive bayes method. They also used color thresholding values for extracting or segmenting the TB bacilli from the back ground image and got an ROC area 99.3%. Rulaningtyas et al. [20] claimed that their color segmentation

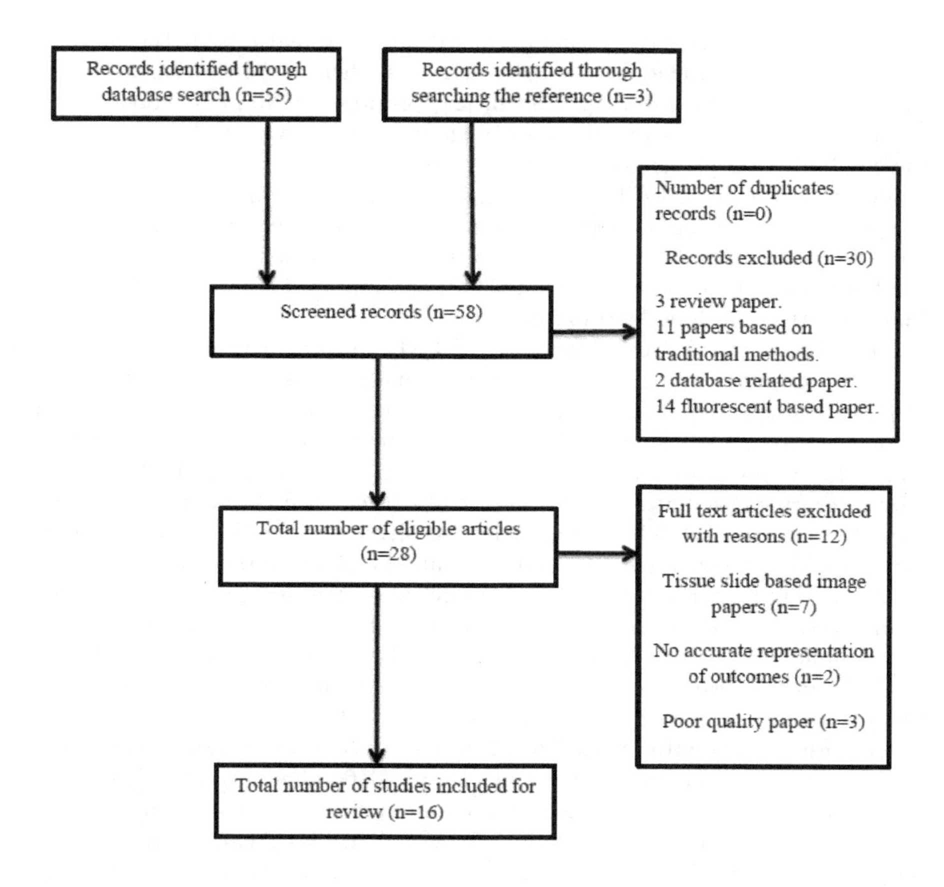

FIGURE 3.2: Study design flowchart for the review of automatic TB detection.

method failed in eliminating all the background images correctly, so it still needs some optimization techniques for increasing the detection accuracy.

(b) Decision tree and random forest based methods In 2010, Zhai et al. [26] presented a fully automatic TB detection system i.e., an auto-focus algorithm and an automatic detection algorithm. They performed a coarse segmentation and fine segmentation based on color information such as HSV and CIE L*a*b*, and finally classified the bacilli and non-bacilli using decision tree method. They build the decision tree based on three shape-based features such as area, roughness and circularity and set a threshold value for each feature (80 for area, 1.2 for roughness, 0.6 for circularity) for the classification. But their method only showed 81% to 90% of accuracy for the majority of test images. The main disadvantages of this method is that TB bacilli cannot be properly extracted based on three features. If the method in Zhai et al. [26] incooperates more features, then the detection accuracy can be improved. In 2014, Ayas et al. [21] proposed a Random Forest (RF) based method for the segmentation and classification. Ayas et al. [21] grouped the candidate TB pixels using connected component analysis and utilized the bounding box facility for bacillus identification. The authors also compared the performance of the proposed method with Support Vector Machine (SVM-sensitivity rate=86.71%) and Neural Network (NN-sensitivity rate=76.82%, total number of hidden layer neuron=300); and reported a better performance than the other machine learning methods.

(c) Support Vector Machine (SVM) based method CostaFilho et al. [3] proposed a bacillus segmentation method based on support vector machine and a post-processing method containing three filters. First step used in their approach is a scalar selection technique for selecting input variables and utilized four color spaces such as RGB, HSI, YCbCr and Lab. The filters (size filter, rule based filter and geometric filter) removed the artifacts and successfully detected the TB bacilli with a sensitivity of 96.8%. However, their method did not considered touching bacilli. Main advantage of CostaFilho et al. [3] was that the authors took more geometric features such as area, compactness, perimeter, eccentricity and Hu moments (first and second order) compared to other studies. The authors also used public dataset of Costa et al. [5] and reported that the best performance with SVM is obtained with the quadratic kernel with C parameter equal to 1.6.

(d) Fuzzy classification based method A fuzzy classification based method was proposed by Ghosh et al. [15] in 2016. The method in [15] has a pre-processing stage, feature extraction stage (shape, color and granularity) and fuzzy classification stage; followed by a final classification stage. They proposed a gradient based region growing technique for accurately detecting the contour of TB bacilli. However, their shape-based method in [15] failed in detecting the overlapping or touching TB bacilli. But they claimed that color- and granularity-based methods detected touching bacilli, so they used a majority voting for the final decision making.

(e) Neural Network based method CostaFilho et al. [4] developed a machine learning technique for TB detection, which was based on feed forward neural network. In this method, the images are classified into either a high density or low density background image using the hue component of the image; and then segmented the bacilli using a set of color features. Also, an additional feature called color ratio was used along with the eccentricity for efficient classification. The authors set a threshold value of 0.77 for eccentricity; and the objects with threshold value greater than 0.77 were considered as bacilli, otherwise considered as noise. The method in CostaFilho et al. [4] got a precision of 99.1% for low density background images and attained a precision of 92.8% for high density background images. However, their method classified the objects as bacillus and non-bacillus and did not mention about touching bacilli. Recently in 2017, Lopez et al. [16] proposed a Convolutional

Neural Network (CNN) model for extracting TB bacilli from sputum smear microscopic images. Their study utilized RGB, R-G and grayscale patch versions as input for three CNN models followed by regularization techniques (normalization, data augmentation and dropout). Experimental analysis in Lopez et al. [16] showed that R-G input version and three CNN layer model with regularization has a superior performance than the other patch versions. One of the main features of CostaFilho et al. [4] and Lopez et al. [16] was that the authors used the same public dataset provided by Costa et al. [5], but we cannot compare the two methods because in Lopez et al. [16], they used TB bacilli patches and in CostaFilho et al. [4], they used full image.

Very recently, Panicker et al. [14] proposed a deep learning method for detecting TB bacilli from sputum smear microscopic images. The authors detected the TB bacilli in two stages, an image binarization stage, which segments the bacilli from sputum smear microscopic images followed by a machine learning based classification stage using convolutional neural network. The main advantage of this approach is that it automatically detect single bacilli and touching bacilli accurately using deep learning techniques. In Panicker et al. [14] the authors compared their results with CostaFilho et al. [3] (both methods used the same dataset) and got a high accuracy with CNN method. However, this method is not good for images with variation in staining (poor background images). A drug sensitive TB detection using deep neural network was proposed by Kant et al. [23] in 2018 and their method adopts a patch wise detection strategy using a five layered Convolutional Neural Network (CNN). But the authors in Kant et al. [23] did not compare their method with other recent related works. Also, no image enhancement technique is applied before giving the images to the CNN model, so the method showed poor detection accuracy.

(f) Hybrid methods A combination of pixel classifiers is utilized by Khutlang et al. [19] and [18] for extracting bacilli objects from conventional sputum smear microscopic images. In Khutlang et al. [19], the authors performed the Fisher transformation based feature extraction and a feature subset selection. Two one-class classifiers were used for bacillus identification in Khutlang et al. [18], i.e., the authors performed one-class pixel classification and one-class object classification. The first step used color information and the second step used shape information; and the performance of each classifier is measured using the Receiver Operating Curve (ROC) method. One of the major drawbacks reported in Khutlang et al. [19] and [18] was their inefficiency in detecting touching TB bacilli; and the authors reported the performance of their method based on single bacilli only. An automatic TB detection system was proposed by Samuel et al. [9] in 2018; and this system is composed of two stages: image capturing stage and bacilli recognition stage. In the second stage, they utilized transfer learning method which was done with the help of customized Inception V3 DeepNet model followed by support vector machine classification. One of the major drawbacks of their method was that it does not support adaptive learning, i.e., it does not work with a poor sensitivity dataset. Very recently in 2018, a Gaussian-Fuzzy-Neural Network (GFNN) based bacilli segmentation technique was presented by Mithra et al. [22]. Here, the authors followed color space transformation, Otsu thresholding and feature extraction techniques; and finally classified the bacilli, non-bacilli and overlapping bacilli by using GFNN. Different features considered for the method [22] are bacilli length, density, area, color, histogram features (mean and variance), and local gradient pattern. The authors used a public dataset (Shah et al. [12]) and got only poor segmentation accuracy. The main disadvantage of their method is that the authors did not compare their results with other recent related works; they also mentioned that the dataset they used is more suitable for mobile-based health applications.

(g) Other machine learning based methods In 2015, Xu et al. [2] employed a marker-based watershed transform method for segmenting TB bacilli from sputum smear microscopic images. The authors performed shape based feature extraction (compactness, eccentricity, Hu moments and roughness) and classified the TB bacilli as single bacilli, touching bacilli and non-bacilli using a logistic regression classifier. The experimental analysis in Xu et al. [2] showed better performance than the other methods like random forest and SVM. A reduced rank based method was suggested by Ayma et al. [1], for segmenting the bacilli objects from sputum smear microscopic images. For that, the authors utilized an adaptive filtering based approach and achieved a recall of 93.56%. From the above studies we can conclude that Machine Learning (ML) techniques contributed significantly in detecting TB disease automatically. Table 3.1 shows the different TB detection studies using machine learning techniques.

TABLE 3.1: Overview of different TB detection methods based on Machine Learning (ML) techniques

Author & year	ML methods used for automatic TB detection	Detection of single bacilli and touching bacilli	Major image processing techniques used for automatic TB detection	Performance measurements
		Bayesian based methods		
Sadaphal et al. [17] (2008)	Bayesian method.	Detected single bacilli; touching bacilli of shape "Y or T" is not detected. Other conglomerations are labelled as "possible-TB" objects.	*Used shape based features such as area, eccentricity, axis length ratio. *Used morphological operations.	Not reported
Rulaningtyas et al. [20] (2012)	Naïve Bayes classifier.	Not considered the touching bacilli	*Pre-processing of the image. *Color segmentation.	Precision: 93.6%

Decision tree and Random forest based methods				
Ayas et al. [21] (2014)	Random forest method.	Classified as bacilli and non-bacilli. Not mentioned about touching bacilli detection.	*Removed the noise in the image by using Mahalanobis distance. *Connected component analysis.	Sensitivity: >89.34% Specificity: >62.89%
Support Vector Machine (SVM) based method				
CostaFilho et al. [3] (2015)	Support Vector Machine.	Detected single bacilli only.	*Used color information for segmentation. *Used color ratio as new parameter.	Sensitivity: 96.8% Error rate: 3.38%
Fuzzy classification based method				
Ghosh et al. [15] (2016)	Fuzzy classification.	Detected single bacilli only. Not detected overlapping bacilli.	*Features used: shape, color and granularity.	Sensitivity: 93.9% Specificity: 88.2%
Neural network based methods				
CostaFilho et al. [4] (2012)	Feed forward neural network.	Classified as bacillus and non-bacillus	*Color ratio and geometric features are considered for classification.	Sensitivity: 91.53% Precision: 91.49%
Lopez et al. [16] (2017)	CNN	Classified as positive patch and negative patch. Not mentioned about touching or overlapping bacilli.	*CNN models are trained by using RGB, R-G and gray scale patches.	ROC Area =99% Accuracy (R-G) =97%
Panicker et al. [14] (2018)	CNN	Detected single bacilli and touching bacilli.	*Fast nonlocal means method for denoising. *Otsu segmentation, morphological operations, connected component analysis.	Recall: 97.13% Precision: 78.4% F-score: 86.76%
Kant et al. [23] (2018)	CNN	Patches are classified as positive, negative and suspect cases.	————	Recall: 83.78% Precision: 67.55%

		Hybrid methods		
Khutlang et al. [19] (2010)	K-nearest neighbour, probabilistic neural networks, SVM, Bayes Classifier.	Detected single bacilli only. Touching bacilli is not detected.	*Features used: Eccentricity, Fourier features, value of the central pixel, and Moment features. *Fisher mapping based dimensionality reduction.	Sensitivity: 97.77% Specificity: 99.13% Accuracy: 98.55%
Khutlang et al. [18] (2010)	K-nearest neighbour, Principal Component Analysis (PCA).	Detected single bacilli only. Touching bacilli is not detected.	*Features used: Moment features, eccentricity, Fourier features	Sensitivity: 94.36% Specificity: 78.65%
Samuel et al. [9] (2018)	Customized Inception V3 DeepNet model and Support Vector Machine.	Classified the input image as TB or not.	*Feature extraction.	Accuracy: 95.05%
Mithra et al. [22] (2018)	Gaussian Fuzzy neural network.	Classified few bacilli, non-bacilli and overlapping bacilli.	*Color space transformation and thresholding.	Segmentation accuracy: 91.379%
		Other machine learning based methods		
Xu et al. [2] (2015)	Logistic regression.	Classified single bacilli, touching bacilli and non-bacilli.	*Used marker based watershed transform. *Shape features: compactness, eccentricity, roughness and Hu moments.	Accuracy: 91.68% Sensitivity: 92.83% Specificity: 87.88%
Ayma et al. [1] (2015)	Least Mean Square and Reduced rank method.	Performed bacilli segmentation only.	*Local Binary pattern descriptor. *Feature vector.	Recall: 93.56%

Nowadays, machine learning technique plays an important role in detecting TB disease automatically. With the help of these advanced technologies, the health care professionals can offer better and accurate quality of treatments to the needed ones. However, different methods of machine learning techniques are adopted by many researchers for TB detection. From Table 3.1, we can understand that most of the studies followed neural network based methods for detecting TB bacilli. Thus machine learning techniques can be considered as an important diagnostic tool for detecting many diseases including TB. Table 3.2 shows the details of dataset and images used by different authors for their study.

TABLE 3.2: Details of datasets and images used for various studies for detecting TB

Author & year	Image details
Private dataset	
Sadaphal et al. [17] (2008)	———-
Zhai et al. [26] (2010)	No. of images used: 100 Image resolution: 1392 x 1040
Rulaningtyas et al. [20] (2012)	No. of data used for training: 2063
Khutlang et al. [19] (2010)	No. of segmented images: 185 Image resolution: 720 x 480
Khutlang et al. [18] (2010)	No. of images: 28 Image resolution: 720 x 480
Ayas et al. [21] (2014)	No. of images used: 116 Image resolution: 640 x 480
Xu et al. [2] (2015)	No. of objects for training: 1000 No. of test image used: 240 Image resolution: 680 x 512
Ayma et al. [1] (2015)	Training samples: 80 Testing samples: 79 Pixel resolution: 684 x 912
Ghosh et al. [15] (2016)	No. of images used: 100
Public dataset (Costa et al. [5])	
CostaFilho et al. [4] (2012)	No. of images used: 120. Image resolution: 2816 x 2112
CostaFilho et al. [3] (2015)	Training samples: 2400 Testing samples: 2456 Pixel resolution: 2816 x 2112
Lopez et al. [16] (2017)	No. of images used: 492 Pixel resolution: 1388 x 1040
Panicker et al. [14] (2018)	Training samples: 1800 Testing samples: 1817 Pixel resolution: 2816 x 2112
Public dataset (Shah et al. [12])	
Mithra et al. [22] (2018)	Pixel resolution: 800 x 600
Kant et al. [23] (2018)	Less no. of images is used.
Private and public dataset (Shah et al. [12])	
Samuel et al. [9] (2018)	No. of images used: 1242 Pixel resolution (private): 1920 x 1280 Pixel resolution (public): 800 x 600

3.4 Discussion

This chapter provides a review of different machine learning techniques used for TB detection. We considered 16 published studies based on conventional sputum smear microscopic images which are based on machine learning techniques, and which are published between 2008 and 2018. As mentioned earlier, sputum smear examination using conventional or fluorescent microscope is the most widely followed cost-effective technique for active TB detection. Nowadays, many hospitals use computerized detection of various diseases like cancer, malaria, etc. which can produce faster and more accurate results than the manual detection method. Computerized detection of TB is also one among them, which can analyse sputum smear images quickly and accurately. Every microscopy-based automatic system consists of auto-focus stage and automatic detection stage. Machine learning techniques can be used in the second stage for accurate detection of all types of TB bacilli. When we analysed the published works from 2008 to 2018, we noticed that only very few researchers [8] and [9] developed fully automatic TB detection systems (most of the researchers developed semi-automatic TB detection systems). Among them "TBDX" is commercially present in the market and its cost is high [13], which cannot be affordable to the majority of the developing countries. So, there is a need for the development of cost-effective systems for TB detection. The main characteristics of an automatic TB detection system are: (a) it improves the sensitivity of the TB test, (b) it reduces the need of highly qualified (and expensive) technicians, and (c) it limits the number of expensive confirmatory tests required. For the treatment of TB disease, periodic examinations of sputum smear microscopic images are required; on using automated systems, the previous images of the patient's sputum smear could be saved for subsequent comparative analysis. Also, these automated systems are very helpful to rural areas, where there is a dearth of skilled technicians. Usage of these types of systems helps to improve outcome and quality of health care. Treatment of TB and its resistant cases (multi-drug, extensive drug and total drug resistant forms) requires the development of more accurate, cost-effective techniques. Many machine learning methods showed better performance in detecting single and touching TB bacilli very accurately than the other conventional methods. From Table 3.1 we can study that the TB detection based on machine learning techniques showed better accuracy and sensitivity (greater than 90%) than the traditional methods. As mentioned earlier, the majority of conventional methods showed low accuracy in detecting TB bacilli, which is accuracy less than 90%. Also, the majority of the conventional methods were good in detecting only single bacilli; but they ignored touching bacilli or considered the touching bacilli as non-bacilli. The methods proposed by Zhai et al. [26], Xu et al. [2] and Panicker et al. [14] successfully classified single bacilli, touching bacilli and non-bacilli. But in Zhai

TABLE 3.3: Comparative analysis of methods based on Costa et al. [5] dataset

Key terms	CostaFilho et al. [4] (2012)	CostaFilho et al. [3] (2015)	Lopez et al. [16] (2017)	Panicker et al. [14] (2018)
Classifier	Feed-forward Neural Network	SVM	CNN	CNN
Performance	91.5%	96.8%	ROC: 99%	97.13%

TABLE 3.4: Comparative analysis of methods based on Shah et al. [12] dataset

Key terms	Mithra et al. [22] (2018)	Kant et al. [23] (2018)
Classifier	Gaussian Fuzzy Neural Network	Convolutional Neural Network
Performance	Segmentation Accuracy=91.379%	Recall: 83.78% Precision: 67.55%

et al. [26] and Xu et al. [2], the authors utilized hand-crafted features for the separation of touching bacilli from single bacilli. In Zhai et al. [26] and Xu et al. [2] the authors designed their methods based on single bacilli features, so it will not work well in detecting all the touching bacilli in a particular image. But in [14], the authors used a deep learning method and successfully separated all types of TB bacilli. Among the considered methods, methods based on CNN are superior to other methods. This is mainly because CNN can improve its performance, if enough training samples are provided. However, unfortunately not many public databases are available for researchers at present. Since in most of the considered papers the results are reported based on private datasets, a fair comparison of the published methods based on the reported results is difficult. However, a comparative analysis of methods on Costa et al. [5] and Shah et al. [12] dataset is given in the Tables 3.3 and 3.4.

From Tables 3.3 and 3.4 we can understand that researchers got very low detection accuracy for the experiments conducted in the dataset of Shah et al. [12] compared to Costa et al. [5]. This conclusion agrees with the findings of Mithra et al. [22], i.e., the dataset [12] is more suitable for mobile based health applications.

3.5 Conclusions and Future Scope

One of the major limitations of this review is that we only considered machine learning techniques based on conventional sputum smear microscopic images. We did not consider the machine learning techniques that are based on fluorescent microscopic images. We ignored fluorescent microscopic images based methods because conventional microscopic images are more popular for TB analysis (especially in developing countries) than fluorescent images. Various challenges involved in the TB detection process are: (1) separation and counting of touching or overlapping TB bacilli, since the severity of the TB disease is always determined through the bacilli count. (2) Building of a cost-effective and user-friendly TB detection system which will work on variation in stained images.

TB is a worldwide epidemic, but ironically it is prevalent in areas where the availability of well-equipped resources like laboratories and expert technicians is very low. In these areas, clinicians and patients do not have access to advanced technologies like automated medical diagnostic systems. So, development of accurate and cost-effective systems like automated systems for TB are very necessary, which will surely help the medical professionals to reduce their clinical burden and also helps to contain the TB burden on society.

Bibliography

[1] Ayma V, De Lamare R, and Castañeda B. An adaptive filtering approach for segmentation of tuberculosis bacteria in Ziehl-Neelsen sputum stained images. In *Latin America Congress on Computational Intelligence (LA-CCI)*, 2015.

[2] Xu C, Zhou D, Zhai Y, and Liu Y. Automatic segmentation and classification of Mycobacterium tuberculosis with conventional light microscopy. In *Proc. of SPIE*, page 9814, 2015.

[3] CostaFilho CFF, Levy P C, Xavier CM, Fujimoto LBM, and Costa MGF. Automatic identification of tuberculosis Mycobacterium. *Res Biomed Eng*, 31(1):33–43, 2015.

[4] CostaFilho C F F, Levy P C Xavier, C M Costa M F, Fujimoto L B M, and Salem J. Mycobacterium tuberculosis recognition with conventional microscopy. In *In Proc. IEEE EngMedBiol Soc.*, page 6263–8, 2012.

[5] Costa M G, Costa Filho C F F, Junior K A, Levy P C, Xavier C M, and Fujimoto L B. A sputum smear microscopy image database for automatic bacilli detection in conventional microscopy. In *In Proc. of 36th Annual International Conference of IEEE Engineering in Medicine and Biology Society (EMBC)*, page 2841–2844, 2014.

[6] Costa M G, Costa Filho C F F, Sena J F, Salem J, and Lima M O. Automatic identification of Mycobacterium tuberculosis with conventional

light microscopy. In *30th Annual International IEEE Eng Med Biol Soc.*, page 382–385, 2008.

[7] `http://apps.who.int/iris/bitstream/handle/10665/274453/9789241565646-eng.pdf?ua=1`, 2018.

[8] Lewis J J, Chihota V N, Meulen M, Fourie P B, Fielding K L, Grant A D, Dorman S E, and Churchyard G J. "proof-of-concept evaluation" of an automated sputum smear microscopy system for tuberculosis diagnosis. *PLoS One*, 7(11):e50173, 2012.

[9] Samuel R D J and Kanna B R. Tuberculosis (TB) detection system using deep neural networks. *Neural Computing and Applications*, pages 1–13, 2018.

[10] Veropoulos K, Campbell C, Learmonth G, Knight B, and Simpson J. The automated identification of tubercle bacilli using image processing and neural computing techniques. In *Artificial Neural Networks*, pages 797–802, 1998.

[11] Sotaquira M, Rueda L, and Narvaez R. Detection and quantification of bacilli and clusters present in sputum smear samples: a novel algorithm for pulmonary tuberculosis diagnosis. In *International Conference on Digital Image Processing*, page 117–121, 2009.

[12] Shah MI, Mishra S, Yadav VK, Chauhan A, Sarkar M, Sharma SK, and Rout C. Ziehl-Neelsen sputum smear microscopy image database: a resource to facilitate automated bacilli detection for tuberculosis diagnosis. *J Med Imaging (Bellingham)*, 4(2):027503, 2017.

[13] Panicker R O, Soman B, Saini G, and Rajan J. A review of automatic methods based on image processing techniques for tuberculosis detection from microscopic sputum smear images. *Journal of Medical Systems*, 40(1):1–13, 2016.

[14] Panicker R O, Kalmady K S, Rajan J, and Sabu M K. Automatic detection of tuberculosis bacilli from microscopic sputum smear images using deep learning methods. *Bio cybernetics and Biomedical Engineering*, 38(3):691–699, 2018.

[15] Ghosh P, Bhattacharjee D, and Nasipuri M. A hybrid approach to diagnosis of tuberculosis from sputum. In *International Conference on Electrical, Electronics, and Optimization Techniques (ICEEOT)*, pages 771–776, 2016.

[16] Lopez Y P, CostaFilho C F F, Aguilera L M R, and Costa M G F. Automatic classification of light field smear microscopy patches using convolutional neural networks for identifying Mycobacterium tuberculosis. In *Proceedings of 2017 CHILEAN Conference on Electrical, Electronics Engineering, Information and Communication Technologies*, 2017.

[17] Sadaphal P, Rao J, Comstock G W, and Beg M F. Image processing techniques for identifying Mycobacterium tuberculosis in Ziehl-Neelsen stains. *International Journal of Tuberculosis and Lung Disease*, 12(5):579–582, 2008.

[18] Khutlang R, Krishnan S, Whitelaw A, and Douglas T S. Automated detection of tuberculosis in Ziehl-Neelsen stained sputum smears using two one-class classifiers. *J. Microsc.*, 237(1):96–102, 2010.

[19] Khutlang R, Krishnan S, Dendere R, Whitelaw A, Veropoulos K, Learmonth G, and Douglas T S. Classification of Mycobacterium tuberculosis in images of ZN-stained sputum smears. *IEEE Trans InfTechnol Biomed*, 14(4):949–957, 2010.

[20] Rulaningtyas R, Suksmono A B, Mengko T L R, and Saptawati. Color segmentation using bayesian method of tuberculosis bacteria images in Ziehl-Neelsen sputum smear. In *In Proc. of WiSE Health*, 2012.

[21] Ayas S and Ekinci M. Random forest-based tuberculosis bacteria classification in images of ZN-stained sputum smear samples. *Signal Image and Video Processing*, 8(1):49–61, 2014.

[22] Mithra K S and Emmanuel W R S. GFNN: Gaussian-fuzzy-neural network for diagnosis of tuberculosis using sputum smear microscopic images. *Journal of King Saud University – Computer and Information Sciences*, page Article in press, 2018.

[23] `https://arxiv.org/pdf/1801.07080.pdf`, 2018.

[24] Makkapati V, Agrawal R, and Acharya R. Segmentation and classification of tuberculosis bacilli from ZN-stained sputum smear images. In *IEEE International Conference on Automation Science and Engineering (CASE)*, page 217–220, 2009.

[25] `http://www.who.int/3by5/TBfactsheet.pdf`, 2018.

[26] Zhai Y, Liu Y, Zhou D, and Liu S. Automatic identification of Mycobacterium tuberculosis from ZN-stained sputum smear: Algorithm and system design. In *Proc. of IEEE International Conference on Robotics and Biomimetics (ROBIO)*, page 41–46, 2010.

Chapter 4

Privacy towards GIS Based Intelligent Tourism Recommender System in Big Data Analytics

Abhaya Kumar Sahoo
Kalinga Institute of Industrial Technology, Deemed to be University, Bhubaneswar, India

Chittaranjan Pradhan
Kalinga Institute of Industrial Technology, Deemed to be University, Bhubaneswar, India

Siddhartha Bhattacharyya
RCC Institute of Information Technology, Kolkata, India

4.1 Introduction 82
4.2 Background 83
 4.2.1 Intelligent Tourism Recommender System and its Basic Concepts 84
 4.2.2 Phases of Tourism Recommender System 84
 4.2.3 Collaborative Filtering Technique Used in TRS 85
  4.2.3.1 Memory-based collaborative filtering 86
  4.2.3.2 Model-based collaborative filtering 87
  4.2.3.3 Evaluation of TRS 88
4.3 Geographical Information System Used in TRS 90
4.4 Big Data Analytics in Tourism 90
4.5 Machine Learning Techniques Used in GIS-based TRS 92
4.6 Privacy Preserving Methods Used in GIS-based TRS 94
4.7 Proposed Privacy Preserving TRS Method Using Collaborative Filtering 97
 4.7.1 Dataset Description 97
 4.7.2 Experimental Result Analysis 97
4.8 Conclusion and Future Work 97
 Bibliography 99

4.1 Introduction

Nowadays, everything is available through the internet. When people are going to buy any kind of product through the internet, they first search for any reviews or comments about that product. At that time, people may be confused about whether that product is preferable or not based on the comments. So, a recommendation system provides a platform to recommend such a product which is valuable and acceptable for people. Such a system is based on features of the item, user preferences and brand information. This filtering based system collects a large amount of information dynamically from the user's interest, ratings, choices or item's behavior, filters this information and provides vital information. The theme of data analytics and big data is not an unfamiliar concept. However, the way it is characterized is continuously varying. Various approaches are made to retrieve large quantities of data efficiently because there are a lot of unstructured and unprocessed data that needs to be processed and can be used in various applications. Nowadays, a voluminous data is produced by various stakeholders, which cannot be handled with traditional data techniques. The Big Data techniques have very promising features to deal with such massive data sets. Data management and deriving useful patterns or conclusions from the data are the main challenges in the big data environment. Big data analytics are applied to structured, semi-structured and unstructured data which is collected from different sources. These analytics can help organization to make better decision for generating high revenue in business.

Internet communication is the main channel used by the tourism industry by which industry can enhance customer operational efficiency, quality of service and customer experiences. Numbers of websites are used to display the details about tourism. Web searching on the internet does not provide more interconnections among selected tourist places. This is the time-consuming process. We cannot get desired objectives through this process. To solve this problem, the tourism recommender system provides right decision making for tourists. With the popularization of internet web-based technologies, social media, smart phones and other hand-held computing devices, the tourism data collection has become easier. Big Data is capable to analyze this data from the tourist industry and make a prediction for further growth of industry. Since Tourism and Big Data have flourished as two immiscible academic disciplines, their cumulative strength is under harnessed.

Tourism Recommender System (TRS) gives more focus on user preferences with free resources and tourist activities of a particular city. TRS requires some useful data which are explicitly collected by users through explicit and implicit feedback. This TRS system is designed and developed based on collaborative filtering techniques. There are mainly two types of recommender systems: user-based and item-based recommenders. Here item denotes tourist location.

In the user-based recommender system, users give their choices and ratings on items. We can recommend that item to the user, which is not rated by that user with the help of user-based recommender engine, considering similarity among the users. In the item-based recommender system, we use similarity between items (not users) to make predictions from users. Data collection for a recommender system is the first job for prediction. Tourism-related data are generated from three primary sources such as users, devices and operations.

As the internet has taken a big role today, users can generate different data like texts, photos and videos, etc. through social media, whereas, due to the development of the internet of things, different sensors can be used to track tourist movements and weather conditions. So, tourism is a kind of system consisting of a series of operations like web searching, site visiting, booking and other activities, etc. which produce transaction data for understanding behavior and interests of tourists along with enhancement of tourism marketing [5][13]. Web-GIS, Social Networking Service (SNS) and recommender system are integrated to develop a GIS (Geographical Information System) which helps in recommending tourist places. In order to support users, useful information about tourist spots can be shared among tourists [4]. When we find the accuracy of a recommender system, at that time we should focus on privacy because our objective is to achieve high accuracy with preserving the privacy of original data in recommender systems. Nowadays, providing privacy to sensitive information in side tourism based recommender system is a main requirement with best accuracy.

The rest of the chapter is organized as follows: Section 4.2 describes the overview of tourism recommender system, its basic concepts and filtering technique used in this system along with its evaluation. Section 4.3 describes geographical information system used in TRS. Section 4.4 tells about big data analytics in tourism. Section 4.5 shows different machine learning techniques used in GIS-based TRS. Section 4.6 shows a comparison among different privacy preserving methods towards GIS based TRS and Section 4.7 presents proposed SVD-based privacy preserving method in TRS. Section 4.8 summarizes this chapter and future work.

4.2 Background

Recommender system requires relevant information which helps to predict and recommend the product in efficient way. Collection of vital information is the main job to build any kind of recommender system. To develop an Intelligent tourism recommender system, preliminary concepts of basic recommender system is required which is discussed in next sections.

4.2.1 Intelligent Tourism Recommender System and its Basic Concepts

The tourism-based recommender system focuses on tourism characteristics and unused resources with customer needs. This system is effective when it automatically learns the user's preferences through collecting explicit and implicit feedback. Implicit interests can be derived through analysis of user's behavior, whereas explicit interests can be inferred in different ways [5].

The intelligent recommender system comprises three layers such as a data source layer, an intermediate layer and an application layer. The server acts in the data source layer which collects data like geographical data and personal information, etc. from different sources. Intermediate layer is the second layer which creates a bridge among data source and application layer. This layer contains business logic related to GIS service applications, recommendation algorithms and network application service. The application layer is the interaction layer by which users can interact with application through different ways [4, 11].

4.2.2 Phases of Tourism Recommender System

To develop an intelligent tourism recommender system, it consists of different phases which are explained below.

- **Information Collection Phase**
  This phase gathers vital information about users and prepares user profiles based on the users' attributes, behaviors or resources accessed by users. Without constructing a well-defined user profile, the recommendation engine cannot work properly. A recommender system is based on inputs which are collected in different ways, such as explicit feedback, implicit feedback, and hybrid feedback. Explicit feedback takes input given by users according to their preferences in a product, whereas in case of implicit feedback obtains user interest indirectly through monitoring user behavior. Combination of both explicit and implicit feedback is known as hybrid feedback [8].

- **Learning Phase**
  This phase takes feedback collected in the previous phase as input and processes this feedback by using a learning algorithm and exploits the user's features as output [8].

- **Prediction/Recommendation Phase**
  Preferable items are recommended for users in this phase. By analyzing feedback collected in the information collection phase, prediction can be made which is happening through model or memory based or observed activities of users by the system [8].

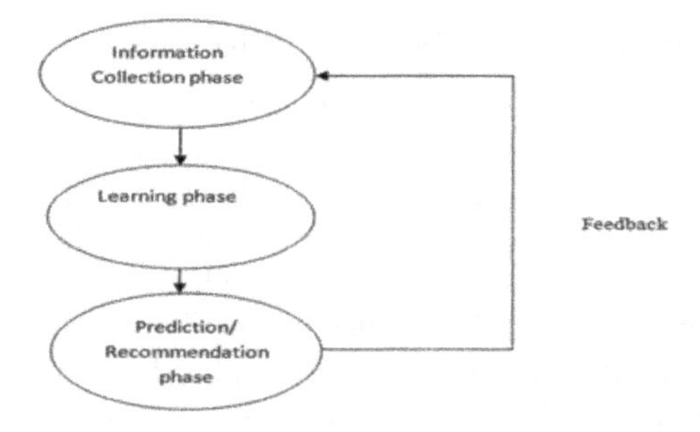

FIGURE 4.1: Phases of the recommender system

4.2.3 Collaborative Filtering Technique Used in TRS

Instead of considering features and attributes of items to determine their similarity, this approach uses user-based ratings to find similarity between items. After collecting all user ratings, the system compares these ratings with other users with the help of a utility matrix and recommends top rated items to the user. We use various distance measures to approach like Jaccard's distance, cosine distance and Pearson's coefficient, etc. to find user similarity. This filtering method is usually used in e-commerce websites to recommend items based on users' ratings.

Collaborative filtering (CF) predicts unknown outcomes by creating a user-item matrix that consists of users' product preferences or interests. Similarities between users' profile are measured by matching the user-item matrix with user preferences and interests. The neighborhood is made of groups of users. The user who has not rated specific items before receives recommendations for a particular product by considering positive ratings given by users in his neighborhood. The CF in recommendation system is used for two purposes: recommendation and prediction. Prediction refers to a rating value Ri,j of item j for user i. There are two types of techniques: memory-based and model-based collaborative filtering. Figure 4.2 explains the whole process of a collaborative filtering technique [8].

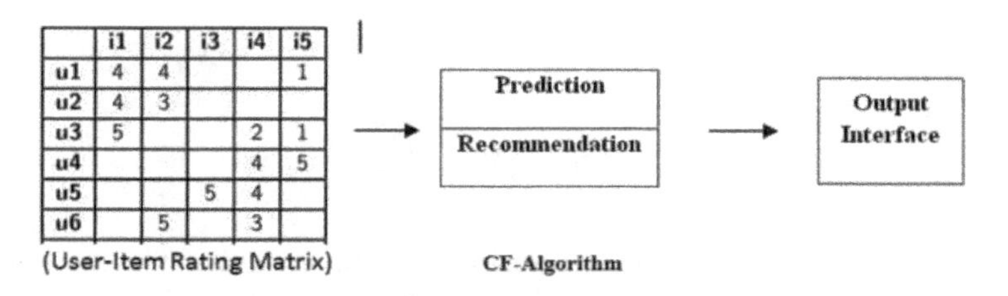

FIGURE 4.2: Collaborative filtering technique

4.2.3.1 Memory-based collaborative filtering

Item and user are two key factors in this filtering technique. So, this technique comprises of item-based and user-based collaborative filtering. Prediction is calculated by measuring similarity among the items. This technique constructs a model on similarities of the item by considering the active user-rated products from the user-item matrix, by which we can measure the similarity among target item and all retrieved items. Then we select k most similar items and find prediction by calculating a weighted average of the active user rating on similar items k. We use different mathematical approaches to calculate user-user and item-item similarity. These are correlation-based similarity measures, cosine-based similarity measures and Pearson's correlation coefficient.

- ***Pearson's Coefficient:*** It can be defined as:

$$P(v,u) = \frac{\sum_{j=1}^{n}(r_{v,j} - \bar{r}_v)(r_{u,j} - \bar{r}_u)}{\sqrt{\sum_{j=1}^{n}(r_{v,j} - \bar{r}_v)^2}\sqrt{\sum_{j=1}^{n}(r_{u,j} - \bar{r}_u)^2}} \tag{4.1}$$

  In the above equation, P(v,u) represents similarity between two users v and u. $r_{v,j}$ and $r_{u,j}$ denote rating on an item i by user v and u, respectively, whereas $\bar{r}_v$ and $\bar{r}_u$ are mean rating given by user v and u respectively, while n is the total number of items in user-item matrix.

- ***Cosine Similarity:*** It can be defined as a vector space model which is based on linear algebra. This method measures similarity between two n-dimensional vectors based on the angle between them. It is mainly used in information retrieval and text mining. The similarity between two items u and v can be denoted by:

$$T(\vec{p},\vec{q}) = \frac{\sum_{j=1}^{n} r_{p,j}\ r_{q,j}}{\sqrt{\sum_{j=1}^{n} r_{p,j}^2}\sqrt{\sum_{j=1}^{n} r_{q,j}^2}} \tag{4.2}$$

- ***Jaccard Similarity:*** Jaccard Similarity of sets P and Q is the ratio of the size of the intersection of P and Q to the size of their union. It can be denoted by:

$$JS(P,Q) = \frac{|\ P \cap Q\ |}{|\ P \cup Q\ |} \tag{4.3}$$

  In a user-based collaborative filtering technique, similarity between users is measured by comparing ratings on the same item. By calculating a weighted average of ratings of item by users, it predicts rating for an item by active user. In this way, the above three methods are used to measure similarity between two items.

4.2.3.2 Model-based collaborative filtering

This approach works on the basis of previous users' ratings to construct a model which uses data mining techniques and applies machine learning sometimes. Different approaches like clustering, association rules, decision tree, regression, artificial neural network and Bayesian classifiers, etc. are used to classify users and items based on the model.

- ***Association Rule Mining:*** Association rule mining algorithms generate association rules which decide the relationship among items in a transaction. Association rule $A \longrightarrow B$ means item set A predicts item set B. This rule can be fitted to a recommendation model to make a prediction about user and item.

- ***Clustering:*** Clustering is a method which is used to partition a set of data into a number of clusters. A good clustering method means high intra-cluster similarity and low inter-cluster similarity. In a recommendation system, users can participate in different clusters partially, and degree of participation can be calculated by taking the average across the clusters of participation [12].

- ***Decision Tree:*** Decision tree makes a graph-like tree structure which is constructed by considering the training data set in which class labels are known. This tree can be used to classify test data. The decision tree is one type of classifier which handles and classifies previous unseen examples.

- ***Artificial neural network (ANN):*** ANN is a network of many connected neurons arranged in different layers. The weight and bias factors are associated with each and every neuron of each layer. Each neuron has a transfer function through which it measures input, process this input and gives output. ANN is a classification technique to classify test data.

- ***Regression:*** Regression is an analysis method where two or more variables are related to each other. One variable is dependent, whereas one or more are independent variables. This regression technique comprises prediction, curve fitting and hypothesis testing, which create relationships among variables.

- ***Bayesian Classifier:*** A Bayesian classifier is used to solve classification problem based on conditional probability and Bayes theorem. Bayesian classifier is used to predict the class through considering the probability of the class with respect to particular attribute by applying Bayes' theorem. This classifier is usually useful when users' preferences change with respect to the time required building the model.

4.2.3.3 Evaluation of TRS

Recommendation systems should be properly evaluated by using different measures. Without evaluation we cannot identify whether our proposed recommendation system is correctly recommending and predicting or not. There are mainly two measures that we consider for evaluation, i.e., prediction accuracy and precision of the recommendation list. The quality of different filtering-based recommendation systems is evaluated based on accuracy and coverage. Accuracy is the ratio of accurate recommendations to total possible recommendations, and coverage can be defined as the percentage of objects (items) in the domain the recommender is able to recommend [8].
Accuracy metrics are used to evaluate the accuracy of any type of filtering-based recommendation system by contrasting the predicted ratings directly with the actual user rating. We use different statistical accuracy metrics such as Mean Absolute Error (MAE) and Root Mean Square Error (RMSE), etc.

- ***Mean Absolute Error (MAE):*** This measure is very easy to understand and widely used for calculating the amount of diversion of recommendation from a user's specific value. It can be defined as:

$$MAE = \frac{1}{N} \sum_{v,j} \mid r_{v,j} - A_{v,j} \mid \tag{4.4}$$

  Where N denotes total number of ratings on the item set, $r_{v,j}$ denotes the predicted rating for user v on the item j, and $A_{v,j}$ is the actual rating. The lower is value of MAE, the better is the accuracy of the recommender.

- ***Root Mean Square Error (RMSE):*** This measure defines standard deviation of the residual errors, i.e., differences between predicted values and known values. RMSE can be defined as:

$$RMSE = \sqrt{\frac{1}{N} \sum_{u,i} (r_{u,i} - T_{u,i})^2} \tag{4.5}$$

  Where N denotes total number of ratings on the item set, $r_{u,i}$ denotes the predicted rating for user u on the item i, and $T_{u,i}$ is the actual rating.

  Recommendation accuracy metrics are used to recommend any product to the user by calculating measurement factors like recall, F-measure and precision, etc. These factors are calculated with the help of a confusion matrix. The confusion matrix is described in Table 4.1.

- Sensitivity is the proportion of actual positives which are correctly identified as positives by the classifier. It is defined as:

$$Sensitivity = \frac{TP}{(TP + FN)} \tag{4.6}$$

TABLE 4.1: Confusion matrix

Total Population	**Condition Positive**	**Condition Negative**
Predicted condition positive	True positive	False positive
Predicted condition negative	False negative	True negative

- Specificity measures the proportion of correctly identifying excluded condition when the condition is not present. This is defined as:

$$Specificity = \frac{TN}{(TN + FP)} \tag{4.7}$$

- Predictive value positive is the proportion of positives that correspond to the presence of the condition. This is defined as:

$$Predictivevaluepositive = \frac{TP}{(TP + FP)} \tag{4.8}$$

- Predictive value negative is the proportion of negatives that correspond to the absence of the condition. It is defined as:

$$Predictivevaluenegative = \frac{TN}{(TN + FN)} \tag{4.9}$$

Precision: This is a measure of retrieved instances that are relevant, while recall can be defined as the fraction of correctly recommended items that are also part of the collection of useful recommended items. Precision is defined as:

$$Precision = \frac{Correctlyrecommendeditems}{Totalrecommendeditems} \tag{4.10}$$

Recall is defined as:

$$Recall = \frac{Correctlyrecommendeditems}{Totalusefulrecommendeditems} \tag{4.11}$$

F-measure aids to combine both precision and recall into a single score or metric. It can be defined as:

$$F - Measure = \frac{2 \times Precision \times Recall}{Precision + Recall} \tag{4.12}$$

Coverage deals with the percentage of items that the recommendation system was able to recommend. If there are many sparse user-rated items, then it is not possible to compute prediction. It can be minimized by defining small neighborhood sizes for users or items. By considering the above parameter measures, we can evaluate any type of recommendation system so that performance can be achieved in terms of accuracy.

4.3 Geographical Information System Used in TRS

Geographical information system (GIS) is a system which is used to capture, store, manipulate and analyze the spatial or geographic data. It is attached to the tourism sector. The GIS is mainly designed using three servers such as database server, web server and GIS server. GIS refers to different processes which relate unrelated information by taking location as a key index variable. Web server works on SNS, accesses the GIS server, whereas database server integrates all the functions [6]. In SNS, the details of system configuration are performed to suit the regional characteristics of a region. In TRS, GIS is used to identify proper tourist locations and feeds information to a recommender system so that TRS keeps track of all information related to different tourist spots and services [4, 7, 16].

4.4 Big Data Analytics in Tourism

The tourism industry is a very productive and promising industry in modern economy, not only from the viewpoint of revenue generation but also to nurture social and cultural relations. In the tourism industry, the customer data are very useful because they provide information about user choice, tour pattern, allied activities and hot tourist destinations, etc. With the popularization of internet web-based technologies, social media, smart phones and other hand-held computing devices, the tourism data collection has become easier. Today, a voluminous data is produced by various stakeholders, which cannot be handled with traditional data techniques. The Big Data techniques have very promising features to deal with such a massive data set. Big Data is able to analyze this data from the tourist industry and make a prediction for further growth of industry [2][13].

There are two phases, named as data collection and data mining, used in tourism research. During the collection phase, we collect online textual data from social media and extract useful hidden information from these data. The data mining phase integrates data processing and pattern discovery sub-phases. In the first phase, online data are collected through web crawling technology. The web crawler is a set of programs used to download web pages, extract URLs from HTML pages and fetch them [15]. In the second phase, the data mining technique is used to extract hidden patterns from collecting a huge amount of textual data through data preprocessing and pattern discovery. Data preprocessing sub-phase, includes data cleaning, tokenization and word streaming approaches applied to the tourism related data, whereas pattern discovery sub-phase, includes latent Dirichlet allocation (LDA), sentiment analysis, statistic analysis, clustering and different regression models used to discover useful patterns [9].

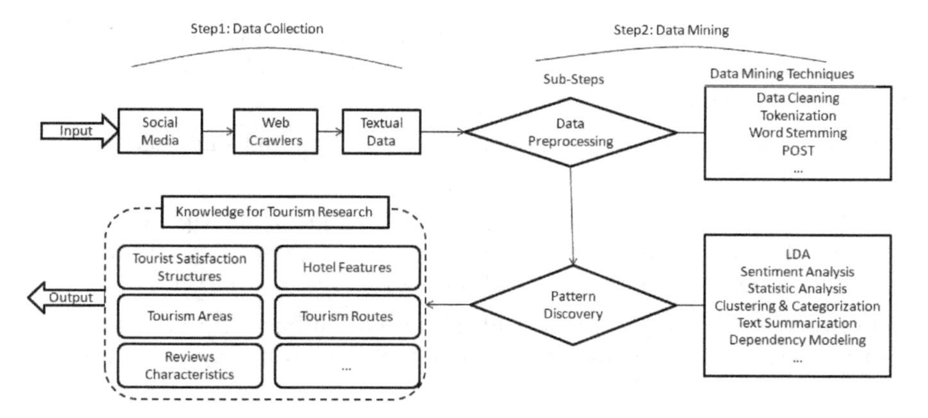

FIGURE 4.3: Steps involved in tourism research

We consider different processes which are most required in intelligent tourism research. These are explained as follows.

- Data cleaning is the main step which removes unwanted data from online textual data. This unwanted data may include low-frequency words, incomplete data and misspellings, etc. which help to create valuable and relevant information about tourism.
- Tokenization is the process by which collected textual data breaks into words, phrases and tokens. By using this method, we get tourism-related keywords like tourism spots and travelling details from a large number of sentences.
- Word stemming is a method to identify the word's root and make one token having the same root for simplicity.
- POST refers to labeling each word with a POS tag. A POS tag includes a noun, adverb and adjective. In this case, unwanted words with other tags are removed.
- Pattern discovery is a main step which discovers useful and interesting patterns from texts by using different techniques such as latent Dirichlet allocation (LDA), sentiment analysis, statistical analysis and clustering, etc.
- The LDA method is used to identify and discover a mixture of abstract topics from collecting reviews.
- Sentiment analysis is the process of opinion mining, which looks into attitudes towards tourism products. This process classifies textual data into sentiment categories such as negative, positive or neutral.

- Statistical analysis is a technique which uses statistical approaches like t-test, correlation analysis to analyze online textual data and produce useful outputs.
- Clustering is used to group similar sets of data and find hidden information from that group.
- Text summarization makes a summary of single or multiple documents by refining collected textual data. Different regression models are used to find relationships among textual data related to tourism.

In this way, we can get useful hidden information from textual data in the tourism sector. Hidden knowledge is definitely helpful in improving tourism management.

4.5 Machine Learning Techniques Used in GIS-based TRS

Different machine learning algorithms like Bayesian, decision tree, matrix factorization, nearest-neighbor, deep learning neural network and clustering are used for prediction of different tourist places according to user's preferences [14].

Bayesian classifier is based on conditional probability. This classifier is applied on training data. Bayes' rule is used to compute the probability of R, given a particular instance of P1,..., Pj and predict the class having the highest posterior probability. That means all attributes of Pi are conditionally independent, given the value of class R. Probabilistic independence is defined as: Pr (P—Q, R) = Pr (P—R) for all possible values of P, Q and R, whenever Pr(R) > 0, where P is independent of Q, given R. This classifier describes about the probability distribution over the training data.

To build a decision tree in recommender systems, the input set is composed of ratings. These ratings are described as a relation ¡ItemID, UserID, Rating¿ in which (ItemID, UserID) is treated as the primary key. The rating is the target attribute which is achieved after classification of attributes by using a decision tree method. Based on the training set, the recommender system predicts and recommends the item to a particular user based on the highest rating value. The decision tree is made through a recursive process. This recursive process starts from root node with an input set (training set). At each node, an item attribute is used as the split attribute.

The basic principle of randomization method is to perturb the data in such a way that the server can only know the range of data. Server should not know the true ratings of each user to a particular item in a recommender system with m users and n items. Tao et al. [10] represented a non-negative matrix factorization (NMF) method to provide privacy to user rating of items. The

NMF method is a mathematical method in which multivariate analysis and linear algebra can be used. An original matrix V is factorized into two matrices, W and H, with the property that all three matrices have no negative elements.

$$v_i = Wh_i \tag{4.13}$$

Where v_i is ith column vector of the product matrix V and h_i is the ith column vector of the matrix H. When multiplying two matrices, the dimensions of the factor matrices should be lower than those of the product matrix and it is the property that forms the basis of NMF.

The K-nearest neighbors method is used to calculate the distance between every interaction vector as data points in the query set and reference set. For each vector in the query set, the 'k' closest reference vectors are found. Here we use two distance measures such as Euclidean distance and cosine similarity. In this approach, we take the average of interaction vectors of 'k' neighbors to get a valid score, which lies between 0 and 1 for each item. Then the predicted interaction scores are used to rank the items.

Deep learning is a sub-branch of machine learning, which learns multiple levels of representations from data. It uses a multiple number of neurons with differentiable objective functions for optimization. Deep neural networks are grouped together to make a single differentiable function. This approach can also be used in recommender systems. Different components are involved in this deep learning approach.

Multilayer Perceptron (MLP), Auto-encoder, Convolutional Neural Network, Recurrent Neural Network, Restricted Boltzmann Machine, Neural Autoregressive Distribution Estimation and Adversarial Networks are the main components of the deep learning method. MLP is a feed-forward neural network with multiple hidden layers in which perceptron uses the arbitrary activation function. Auto-encoder is an unsupervised model which attempts to reconstruct its input data in the output layer. Convolutional Neural Network is a special kind of feed-forward neural network with convolution layers which captures the global and local features for enhancing the efficiency and accuracy. Recurrent Neural Network is used for modeling sequential data. Restricted Boltzmann Machine is a two-layer neural network, which consists of visible layer and hidden layer. There is no intralayer communication among visible and hidden layer. Neural Autoregressive Distribution Estimation is an unsupervised neural network with autoregressive model which is composed of feed-forward neural networks. An Adversarial Network is a generative neural network, which consists of a discriminator and a generator. The two neural networks are trained simultaneously by competing with each other in a minimax game framework.

Clustering method makes a number of user clusters to produce recommendations for active users. This method uses 'm' number of users' points working on 'n' number of items. K-means clustering algorithm is used to categorize users based on their interests and make different groups based on similar points.

4.6 Privacy Preserving Methods Used in GIS-based TRS

There are different privacy preserving techniques used in collaborative filtering based tourism recommender system to provide security to sensitive information related to user and item. The comparison study among different techniques is shown in Table 4.2.

- ***Privacy preserving naive Bayesian classification with random perturbation method:***

  Each user u_i divides items into M groups and perturbs the ratings of every group 'n' individually. This method comprises two preprocessing methods. The first method forms a neighborhood based on binary similarity measure. The second method chooses empty cells randomly and fills them with personalized ratings. If common-rated items are given, then we use the Tanimoto coefficient method [3]. Let $R_{i,j}$ be the number of occurrences of commonly rated items with i in the first pattern and j in the second pattern, where $i, j \in 0, 1$. Given two binary feature vectors X and Y, T(X, Y) denotes modified Tanimoto coefficient between X and Y and it can be calculated as:

  $$T(X,Y) = \frac{(r_{1,1} + r_{0,0}) - (r_{1,0} + r_{0,1})}{r_{1,1} + r_{0,0} + r_{1,0} + r_{0,1}} \tag{4.14}$$

  Where $R_{1,1}$ is the number of users who have rated both items as 1. $R_{1,0}$ represents the number of users rated item i as 1 and item j as 0. $R_{0,1}$ is the number of users rated item i as 0 and item j as 1 and $R_{0,0}$ shows the number of users rated both items as 0. Tanimoto similarity measure computes the similarity between two binary vectors. This method is used to recommend best tourist places to users by considering Tanimoto coefficient as distance measure and also provide privacy towards a tourist dataset.

- ***Non-negative matrix factorization method:***

  The concept of a randomization method is to change the data in such a manner that the server can get the idea about the range of data, not the original data. The server should not know about true ratings of users to particular items in recommender systems. Tao et al. proposed non-negative matrix factorization method (NMF) for preserving privacy to user rating to items. The NMF method is purely based on linear algebra and multivariate analysis where an original matrix V is divided into two sub-matrices W and H, in such a way that all three matrices have no negative elements [10].

  $$v_i = Wh_i \tag{4.15}$$

Where v_i is ith column vector of the product matrix V and h_i is the ith column vector of the matrix H. When multiplying two matrices, the dimensions of the factor matrices should be lower than those of the product matrix and it is the property that forms the basis of NMF. NMF generates factors with significantly reduced dimensions compared to the original matrix. For example, if V is an $m \times n$ matrix, W is an $m \times p$ matrix, and H is a $p \times n$ matrix, then p can be significantly less than both m and n. Finally, each user sends this value to the server where the user-item matrix R is created. The matrix R is factorized into W and H using NMF, the rating of user i for item j can be predicted as:

$$P_{i,j} = \bar{r}_i + sgn(R\colon (i,j)).\sigma_i.[W(i).H(j)] \tag{4.16}$$

Where r_i and σ_i are mean value and standard deviation of user i respectively; $sgn(R\colon (i,j))$ is the signature of element in the row i and column j of the R'. This method provides privacy of the sensitive data by factorizing original tourism data input matrix.

- ***Modified random perturbation method:*** This method is based on rating privacy which focuses on privacy of personal ratings. A perturbation range is created for each rating, by which actual rating information is not known to others. Before sending actual rating to the server, random values are added to the original ratings. This randomization technique is used for privacy preserving in recommender systems. The advantage of using this perturbation method is that it is easily installed on the client side [1].

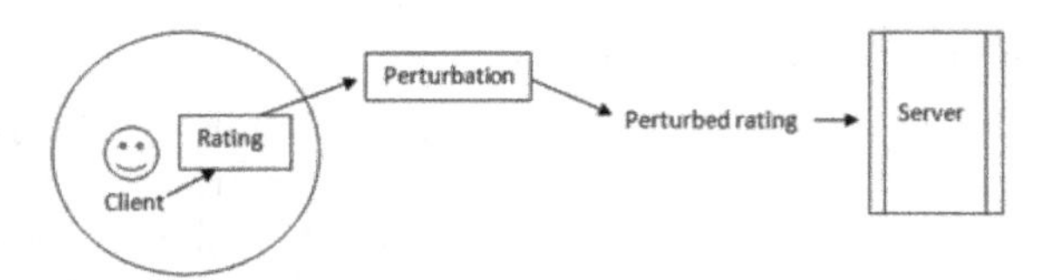

FIGURE 4.4: Randomization method using modified ratings

In the randomization method, use of different range of random values with the original ratings protects privacy in recommender systems. The recommender system uses a rating scale where maximum rating value is 5 and minimum rating value is 1. If the value lies below minimum rating value, then perturbed rating takes a minimum rating value. If it exceeds the maximum rating value, then perturbed rating takes a maximum rating value. This method is used to preserve privacy ratings in tourism-based recommender system.

TABLE 4.2: Comparison among different methods in security and privacy in CF-based recommendation system [Movielens-1M Dataset]

Privacy Preserving Algorithms on TRS	**Description**
Naive Bayesian classification method with random perturbation technique	• Due to its given importance to commonly rated items, Tanimoto coefficient is preferable. • Two preprocessing methods such as neighborhood formation and increasing density to perturbed ratings are used for better privacy and enhancement of accuracy.
Non-negative matrix factorization method	• MAE measures the prediction veracity by calculating the difference between the predicted ratings and the real ratings, so the smaller MAE, the better recommendation quality. • MAE=0.7 for 100 users
Modified random perturbation method	• This is multi-level privacy preserving method by perturbing each rating before it is submitted to the server. The perturbation method is based on multiple levels and different ranges of random values for each level. • Before the submission of each rating, the privacy level and the perturbation range are selected randomly from a fixed range of privacy levels. • MAE=1.3 for k=20 (k=no. of nearest neighbors)

4.7 Proposed Privacy Preserving TRS Method Using Collaborative Filtering

The proposed algorithm is divided into two steps: (1) the original data is disturbed by the method of singular value decomposition (SVD); (2) the additional disturbance is added on the data. The original data matrix S is decomposed by the SVD algorithm, and three factor matrices P, Σ and Q are obtained. Select the parameter k, and get the disturbed matrix. Use the Pearson correlation measure formula to calculate the similarity matrix, then calculate prediction.

4.7.1 Dataset Description

Here we use the US Tourism 2015 dataset which is applied on proposed intelligent TRS. This tourism dataset contains 24 attributes as tourism-related outputs and 27 attributes as tourism-related employment. This dataset is divided into training and test data in 75:25 ratios, respectively. Here 10-fold cross-validation scheme is used while evaluating the results. Our proposed TRS is designed and tested on the tourism dataset which describes rating information along with details. The experimental result is described in Section 7.2.

4.7.2 Experimental Result Analysis

Here we compared the results in terms of mean absolute error (MAE) value among existing methods and proposed TRS by analyzing the tourism dataset. As we get less MAE value in case of our proposed approach, we can say that our approach can be more useful in a healthcare recommendation system.

In Table 4.3, MAE values are shown for tourism dataset, where different privacy preserving methods along with proposed methods are applied. Figure 4.5 depicts that MAE value is lower when proposed method is used. The lower the MAE value, the higher the accuracy. By using collaborative-based filtering technique on proposed TRS, we get less MAE values by achieving high accuracy as compared to existing approaches.

4.8 Conclusion and Future Work

In this chapter, we propose an SVD-based privacy preserving method for tourism-based recommender system that works on the basics of big data analytics along with geographical information systems. This method gives better

TABLE 4.3: Comparison among existing approaches and proposed SVD method

Privacy Preserving Methods	**Average MAE Values (Tourism-related output)**	**Average MAE Values (Tourism-related employment)**
Naive Bayesian classification with random perturbation method	0.732	0.793
Non-negative matrix factorization method	0.721	0.781
Modified random perturbation method	0.714	0.776
Proposed SVD method	0.685	0.746

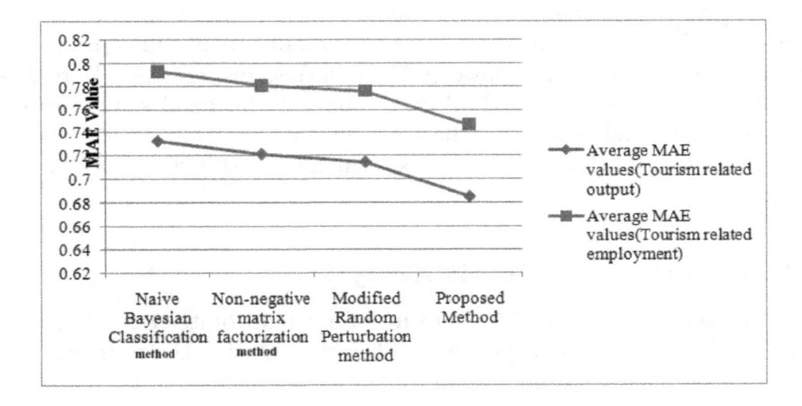

FIGURE 4.5: Comparison among MAE and different privacy preserving methods.

accuracy while preserving privacy than existing methods in TRS. It is able to predict best profit for tourists in different contexts and retrieve data about unknown tourist places which help tourists. With the input of big data analysis, better solutions to current problems related to tourism can be achieved. The trend of acceptance of recommender systems for tourism from researchers, tourists and administrators is increasing steadily with the improvement of tourism systems which helps tourists frequenting different tourist places. We will try to overcome the disadvantages of proposed TRS towards privacy preserving in future work.

Bibliography

[1] Zeynep Batmaz and Huseyin Polat. Randomization-based privacy-preserving frameworks for collaborative filtering. *Procedia Computer Science*, 96:33–42, 2016.

[2] Rajat Kumar Behera, Abhaya Kumar Sahoo, and Chittaranjan Pradhan. Big data analytics in real time-technical challenges and its solutions. In *2017 International Conference on Information Technology (ICIT)*, pages 30–35. IEEE, 2017.

[3] Alper Bilge and Huseyin Polat. Improving privacy-preserving nbc-based recommendations by preprocessing. In *Web Intelligence and Intelligent Agent Technology (WI-IAT), 2010 IEEE/WIC/ACM International Conference on*, volume 1, pages 143–147. IEEE, 2010.

[4] Yolanda Blanco-Fernandez, Martin Lopez-Nores, José J. Pazos-Arias, Alberto Gil-Solla, and Manuel Ramos-Cabrer. Exploiting digital TV users' preferences in a tourism recommender system based on semantic reasoning. *IEEE Transactions on Consumer Electronics*, 56(2):904–912, 2010.

[5] Jesús Bobadilla, Fernando Ortega, Antonio Hernando, and Abraham Gutiérrez. Recommender systems survey. *Knowledge-Based Systems*, 46:109–132, 2013.

[6] Michael N. DeMers. *Fundamentals of Geographic Information Systems.* John Wiley & Sons, 2008.

[7] Tsukasa Ikeda and Kayoko Yamamoto. Development of social recommendation gis for tourist spots. *Development*, 5(12):8–21, 2014.

[8] FO Isinkaye, YO Folajimi, and BA Ojokoh. Recommendation systems: Principles, methods and evaluation. *Egyptian Informatics Journal*, 16(3):261–273, 2015.

[9] Jingjing Li, Lizhi Xu, Ling Tang, Shouyang Wang, and Ling Li. Big data in tourism research: A literature review. *Tourism Management*, 68:301–323, 2018.

[10] Tao Li, Chao Gao, and Jinglin Du. A nmf-based privacy-preserving recommendation algorithm. In *2009 First International Conference on Information Science and Engineering*, pages 754–757. IEEE, 2009.

[11] Hai-ling Liu, Jun-huai Li, and Jun Peng. A novel recommendation system for the personalized smart tourism route: Design and implementation. In *Cognitive Informatics & Cognitive Computing (ICCI* CC), 2015 IEEE 14th International Conference on*, pages 291–296. IEEE, 2015.

[12] Xiao Ma, Hongwei Lu, Zaobin Gan, and Jiangfeng Zeng. An explicit trust and distrust clustering based collaborative filtering recommendation approach. *Electronic Commerce Research and Applications*, 25:29–39, 2017.

[13] Zhang Mu, Chen Shan, Luo Jing, and Feng Lei. Design of the tourism-information-service-oriented collaborative filtering recommendation algorithm. In *Computer Application and System Modeling (ICCASM), 2010 International Conference on*, volume 13, pages V13–361. IEEE, 2010.

[14] Ivens Portugal, Paulo Alencar, and Donald Cowan. The use of machine learning algorithms in recommender systems: a systematic review. *Expert Systems with Applications*, 2017.

[15] Sheoran, S. K. Big data: A big boon for tourism sector. *International Journal of Research in Advanced Engineering and Technology*, 3:10–13, 2017.

[16] Peng Yue and Liangcun Jiang. Biggis: How big data can shape next-generation gis. In *Agro-geoinformatics (Agro-geoinformatics 2014), Third International Conference on*, pages 1–6. IEEE, 2014.

Chapter 5

Application of Artificial Neural Network: A Case Study of Biomedical Alloy

Amit Aherwar
Mechanical Engineering Department, Madhav Institute of Technology and Science, Gwalior, M.P., India

Amar Patnaik
Mechanical Engineering Department, Malaviya National Institute of Technology, Jaipur, Rajasthan, India

5.1 Introduction 102
5.2 Test Material and Methods 105
5.2.1 Test Materials 105
5.2.2 Manufacturing of Orthopaedic Material 105
5.2.3 Material Characterization 107
5.2.4 Mechanical Studies 107
5.2.5 Wear Measurement of Orthopaedic Material 107
5.2.6 Taguchi Design of the Experiment 109
5.3 Results and Discussions 110
5.3.1 Phase Analysis and Microstructure 110
5.3.2 Mechanical Studies of Manufactured Material 111
5.3.2.1 Micro-hardness 111
5.3.2.2 Compressive strength 113
5.3.3 Taguchi Experimental Design 113
5.4 Simulation Model for Wear Response 114
5.4.1 Data Processing in ANN Model 116
5.4.2 Network Training 117
5.4.3 Neural Network Architecture 120
5.4.4 ANN Prediction and its Factor 120
5.5 Conclusion 124
Bibliography 125

5.1 Introduction

Today, the major problem faced by doctors and biomaterials engineers is the selection of the right biomaterials with correct weight compositions. Metallic biomaterials such as SS, CoCr alloys, and Ti alloys are the effective orthopaedic materials used for implants. Table 5.1 lists the various materials including metallic, composites, ceramic, and so on used in orthopaedic application [8, 13] and Table 5.2 lists the various possible material combinations of orthopaedic implants [38, 55]. These combinations of metallic biomaterials as listed in Table 5.2 have good corrosion resistance, good mechanical properties and biocompatibility, which make them a splendid choice for clinical applications [6]. However, instead of these properties, there is some flaw in the metallic orthopaedic materials such as their corrosive nature. Corrosion deteriorates the implant materials in the form of metal ions and these ions liberated into the tissue resulting in adverse reactions [16]. 316 and 316L grades of stainless steel are the prime grades utilized to manufacture artificial bones as it is easy to cast into distinct shapes and sizes. Grade 316L has a healthier corrosion resistance as compared to 316 grades due to the attendance of less percentage of carbon content in the matrix alloy. Both 316 and 316L grades of stainless steel are easily made into fracture plates, screws, and hip nails. Due to ease of the manufacturing process of stainless steel, it becomes the predominant implant alloy [5]. Based on the superior results, the ASTM have strongly recommended 316L grade as a foremost alloy for implant production [3, 33]. However, one more metallic material such as cobalt-based alloys is amid the most favorable orthopaedic materials for making implant components such as hip and knee joints, owing to its excellent mechanical strength and corrosion resistance [30]. Co-Cr alloy is commonly utilized as implants and fixations due to being much attuned with human body. In contrast with further biomaterials, Co-Cr alloys have a good biocompatibility property than SS (Both 316 and 316L) alloys, but less than titanium alloys. Basically, Co-based alloys can be cast, wrought, or forged [19, 24]. Accumulation of nickel in the modified Co-30Cr-4Mo alloy enhances the corrosion properties due to slow rate oxidation and improved mechanical properties [27, 31, 32]. The density of pure nickel is 8.89 g/cm^3; melting temperature is 1468°C and has an elastic modulus of 209 GPa. Nickel specifically stabilizes the FCC structure and gives strength of the modified alloy. In fact, nickel also improves the workability and castability of the alloy [36]. The main cause of failure of orthopaedic material for hip joint is looseness between the contacting surfaces due to wear and friction. Hence, wear is also a leading factor to be pointed out [15, 58].

The authors in their earlier work [5] fabricated modified Co-Cr alloy (i.e., Co-30Cr-4Mo) with concentration of nickel for hip implants and optimized it. From their results, it was concluded that the 1wt.% nickel concentration alloy provides minimal material loss. This chapter presents how neural network computation technique helps to measure the tribological properties via

TABLE 5.1: Various materials used in orthopaedic application

Materials		**Applications**	**Ref.**
Metals	SS-316L	Used in femoral head and stem components	[5]
	Co alloys	Used in femoral heads and stems, Porous coatings, tibial and femoral components	[19]
	Cast CoCrMo	Used in femoral heads and stems, Porous coatings, tibial and femoral components	[24]
Titanium-based materials	CP Ti	Used in porous coatings in ceramic composites	[33]
	Ti6Al4V	Used in femoral heads and stems, Porous coatings, tibial and femoral components	
	Ti5Al2.5Fe	Used in femoral head and stem components	
	Ti-Al-Nb	Used in femoral head and stem components	
Ceramics	Carbon	Used in metal coatings on femoral stem components and acetabular cup components	[21, 41, 56]
	Alumina	Used in metal coatings on femoral stem components and acetabular cup components	
	Zirconia	Used in metal coatings on femoral stem components and acetabular cup components	
Polymers	PMMA	Used in acetabular cups and patellar parts	[14]
	UHMWPE/ HDPE	Used in porous coatings on metallic and ceramic femoral stem components	
	Polysuffolene	Used in porous coatings on metallic and ceramic femoral stem components	
	PTFE	Used in porous coatings on metallic and ceramic femoral stem components	

prediction method to avoid experimental test, time and costs. The application of the neural network approach on the orthopaedic material has not been reported yet. Hence, this study has a contribution to the allied literature in this concern. ANN modelling in the meadow of tribology has been observed by numerous authors [28, 47, 50, 61]. Rout and Satapathy applied a neural network model to predict the influence of rice husk on wear resistance potential of epoxy resin [46]. The ANNs are also applied in material engineering and

FIGURE 5.1: Scanning electron microscopy micrograph of (a) cobalt; (b) chromium; (c) molybdenum; and (d) nickel.

machining tool of conventional engineering materials. ANN can often solve several problems quicker from the other techniques with the additional ability to learn. Many researchers implemented the ANN to predict the tribological characteristics of MMCs also [17, 18, 57]. Gyurova and Friedrich [23] implemented the ANN to predict the friction and wear measurement of polyphenylene sulphide matrix composites by means of pin-on-disc tribometer. Altinkok and Koker [7] studied the prediction of mechanical strength of MMC with the help of ANN. The use of NN to predict the tribological examination of diverse material/mechanical systems was accounted by Senthil et al. [50]. Prediction of wear properties of A356/SiCp Metal Matrix Composites using ANN was examined by Rashed and Mahmoud [43]. One more objective of this study is to develop the feed forward (FF) multilayer perceptron (MLP) ANN model trained with back propagation (BP) routine and apply this model to predict the tribological properties of nickel filled Co-30Cr-4Mo alloy. The inputs into the network model are filler (i.e., nickel), sliding distance, sliding velocity and normal load. However, an ANN model has not yet been formed earlier for modified Co-Cr-Mo biomedical alloys with distinct weight percentage of nickel produced by specially vacuum casting furnace. Therefore, the idea of this study is to predict the tribology characteristics of modified Co-30Cr-4Mo orthopaedic biomaterials with the use of ANN.

TABLE 5.2: Materials combination used in orthopaedic hip implants

Material Used for Femoral-Socket Component	Results	Reference
CoCrMo-CoCrMo	Aseptic loosening rate is high Restricted use Minimum wear rate	[26, 29, 42, 59]
CoCrMo-UHMWPE	Widely in use Minimum wear loss	[29, 42, 59]
Alumina/Zirconia-UHMWPE	Minimum wear rate	[21, 33, 41, 56]
Alumina-Alumina	Low wear rate Problem of joint pain	[33, 56]
Ti6Al4V-UHMWPE	Maximum wear	[14]
Surface-coated Ti6Al4V- UHMWPE	Better wear resistance to abrasion Attained skinny layer	[14]

5.2 Test Material and Methods

5.2.1 Test Materials

Mittal Industries, India supplied the grades of raw materials such as cobalt (Co), chromium (Cr), molybdenum (Mo) and nickel (Ni) to make test specimens with distinct wt.% Ni, in which cobalt was in ingot form and the rest were in powder form with size below 44 μm. The micrographs of all the alloying elements are shown in Figure 5.1. Furthermore, the element content with weight percentage of the manufactured alloys is listed in Table 5.3.

5.2.2 Manufacturing of Orthopaedic Material

The schematic view of vacuum based casting setup used for manufacturing of orthopaedic materials is shown in Figure 5.2. In the presented work, five plates (100mm x 65mm x 10mm) with x wt.% of Ni (x = 0, 1, 2, 3 and 4) added Co-30Cr-4Mo alloy were produced using an induction furnace according to the composition. In this apparatus, there are two separate sections available in the setup: (1) melting section and (2) casting section. A chiller unit is also attached in the setup. Here, both sections are under vacuum background. A motor was coupled with the bottom section, and set to be around 200rpm for proper mixing of all the metals present in the matrix. For manufacturing, all the proposed materials with respective weight percentages were melted above 1800°C for 12mins and then dropped downward into the graphite mold (100mm x 65mm x 10mm) with the help of plunger under vacuum conditions.

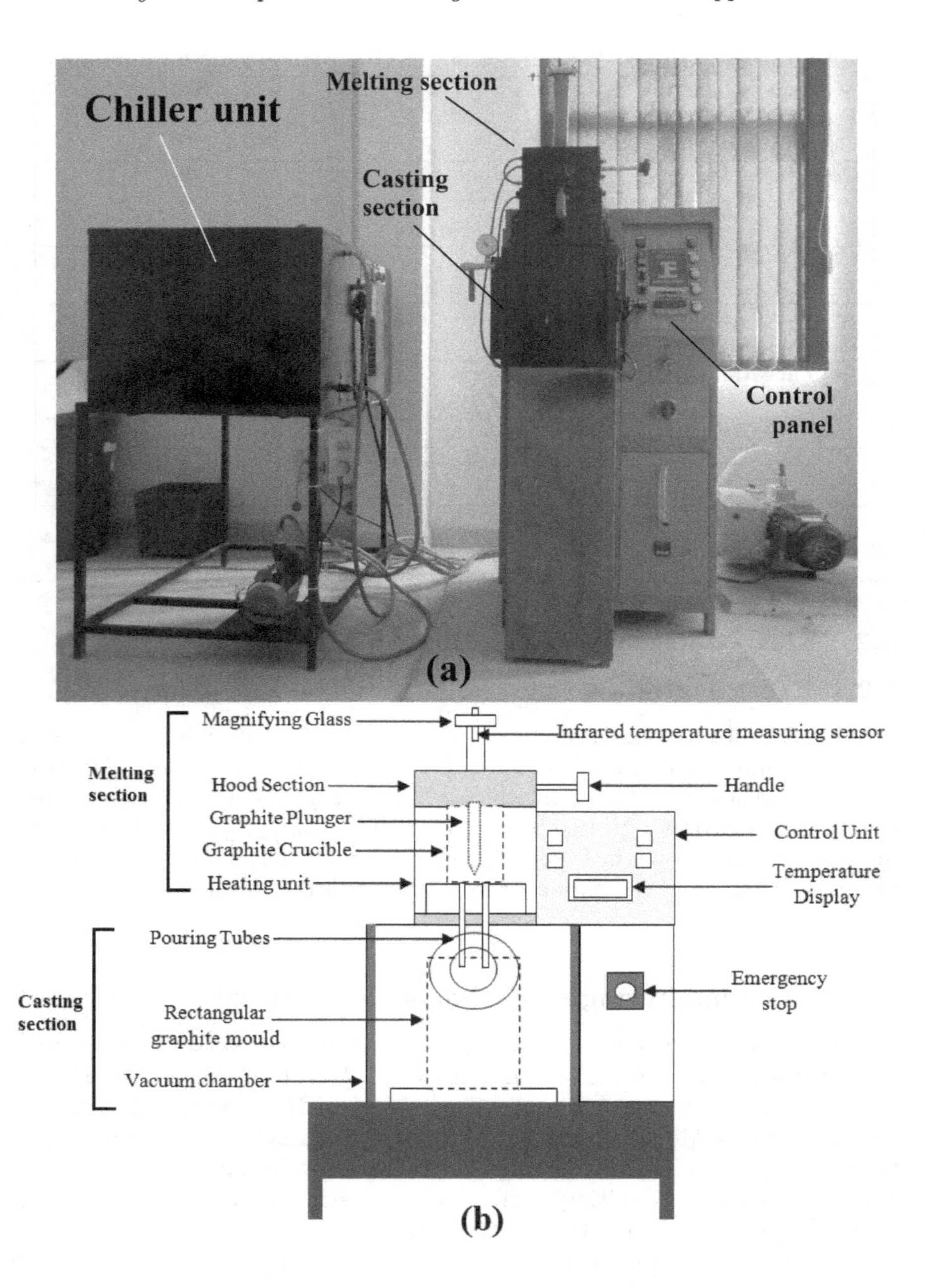

FIGURE 5.2: Experimental setup of casting machine. (a) Image of the casting setup and (b) Schematic diagram of the casting machine used for making the specimens.

TABLE 5.3: Element content and designation of specimens (in weight percentage)

Sample Designation	Elements			
	Co	Cr	Mo	Ni
N0	Bal	30wt.%	4wt.%	0wt.%
N1	Bal	30wt.%	4wt.%	1wt.%
N2	Bal	30wt.%	4wt.%	2wt.%
N3	Bal	30wt.%	4wt.%	3wt.%
N4	Bal	30wt.%	4wt.%	4wt.%

After casting, the mold was removed and then cut as per the sample size.

5.2.3 Material Characterization

The morphology of materials was characterized using FESEM and EDS of FEI Nova Nano SEM 450. The phase and crystal structure of the manufactured orthopaedic materials were studied using XRD and for the same PANalytical X'Pert PRO X-Ray Diffractometer was utilized. The fabricated samples were polished by using Buehler MetaServ 250 polisher/grinder. After polishing, the specimens were etched for 12 sec.

5.2.4 Mechanical Studies

Micro-hardness evaluation was carried out by USL micro-hardness tester. A diamond indenter was indented into the manufactured specimen under a load 0.5N for 10s. The density of the manufactured alloys was recorded by the Archimedian principle. Compression test was carried out on AIMIL machine at a crosshead speed of 2mm/min. For measurement of physical and mechanical properties, three specimens were tested and the mean value was recorded for more precise results.

5.2.5 Wear Measurement of Orthopaedic Material

DUCOM, India supplied a pin-on-disc (POD) tribometer (see Figure 5.3a) confirming to ASTM G 99 [10] was used to conduct the experiments. During the experiment, the specimen (Pin shape; 10mm x 10mm x 20mm) was held stationary while the disc (EN-31 alloy steel of diameter 165 mm x 8 mm thick) was rotated (see Figure 5.3b) and the load was applied through a lever mechanism. The pin samples were cleaned with acetone before executing the wear tests. All the tests were reiterated thrice (Table 5.4) and the average value was taken individually for more precision. The volumetric wear loss (mm3)

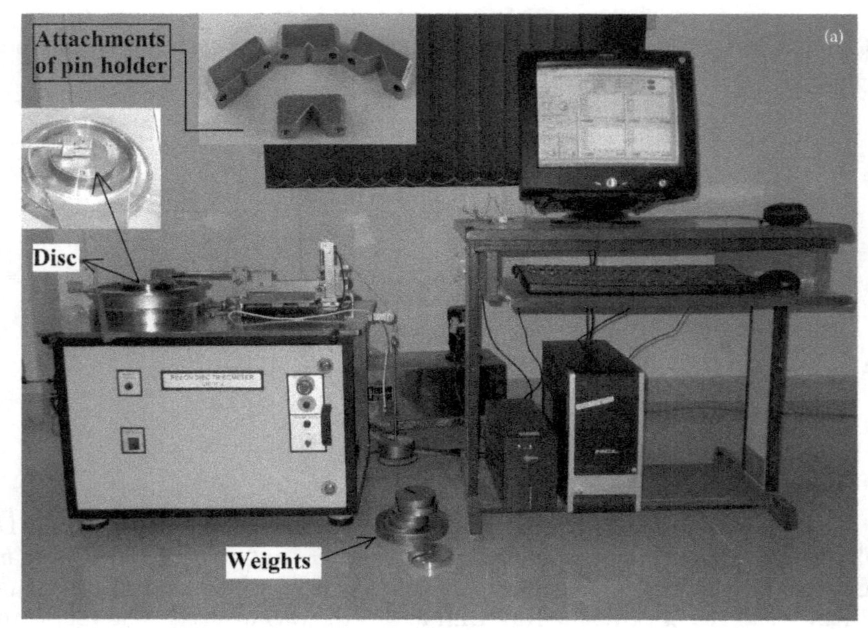

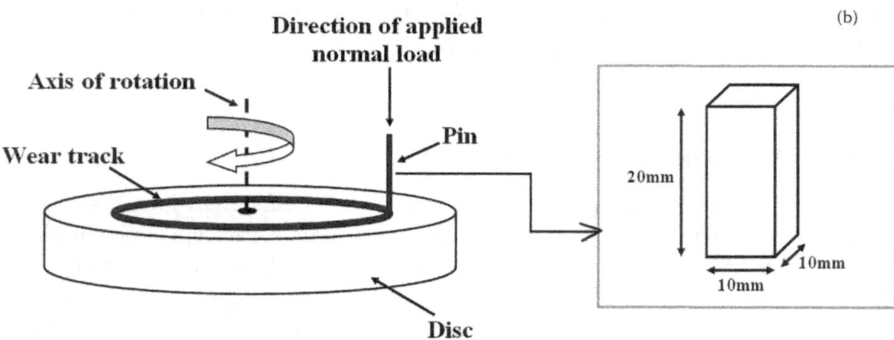

FIGURE 5.3: Test setup (a) pin-on-disc machine and (b) closed view of rotating disc and pin sample.

was computed by Eq. 5.1.

$$Volumtericwearloss = \frac{\Delta m}{\rho_n} \tag{5.1}$$

Where Δm is mass loss in grams, and ρ_n is the density of pin specimen (g/mm^3).

TABLE 5.4: Parameter settings and their levels used in the experiment

Design Factors	**Description**	**Units**	**Level**				
			1	**2**	**3**	**4**	**5**
A	Normal load	N	5	10	15	20	25
B	Sliding velocity	m/s	0.26	0.52	0.78	1.04	1.3
C	Sliding distance	m	500	1000	1500	2000	2500
D	Filler content	wt.%	0	1	2	3	4
Fixed Parameters							
Disc	EN-31						
Test temperature	28°C						
Running time	10 min						
Wear track diameter	50 mm						
Surface roughness	1.6 μm						
Repeated test for each sample	03						

5.2.6 Taguchi Design of the Experiment

Taguchi method helps for exploring the effect of the control factors on response. This method is frequently utilized because of it is simple, systematic and efficient. This method has been applied earlier to evaluate the wear measurement of different composites by Aherwar et al. [2] and Patnaik et al. [40]. Taguchi method mainly diminishes the number of experiments without data loss and is a powerful tool to resolve the difficult problem easily [11, 12, 35, 37, 52, 53]. Taguchi technique formulates a standard orthogonal array (OA) and set the plan of experiments [4, 60]. In this chapter, L_{25} OA was utilized (Table 5.5). The analysis of test data was prepared DOE software known as MINITAB 16. The control factors and their levels considering for testing wear measurement are presented in Table 5.4. The test results obtained from the experimentation were converted into signal-to-noise (S/N) ratio. For this, the smaller the better characteristics are implemented for the analysis. The S/N ratio can be expressed as given below [3].

$$\frac{S}{N} = -10log\left[\frac{1}{k}\sum_{i=1}^{k} {y_i}^2\right] \tag{5.2}$$

Where y_i is the observed data, k is the number of tests.

TABLE 5.5: Taguchi L_{25} orthogonal array design

Trials	Load (N)	Sliding Velocity (m/s)	Sliding Distance (m)	Filler Content (wt.%)
1	1	1	1	1
2	1	2	2	2
3	1	3	3	3
4	1	4	4	4
5	1	5	5	5
6	2	1	2	3
7	2	2	3	4
8	2	3	4	5
9	2	4	5	1
10	2	5	1	2
11	3	1	3	5
12	3	2	4	1
13	3	3	5	2
14	3	4	1	3
15	3	5	2	4
16	4	1	4	2
17	4	2	5	3
18	4	3	1	4
19	4	4	2	5
20	4	5	3	1
21	5	1	5	4
22	5	2	1	5
23	5	3	2	1
24	5	4	3	2
25	5	5	4	3

5.3 Results and Discussions

5.3.1 Phase Analysis and Microstructure

The XRD patterns of all the alloys are presented in Figure 5.4. The microstructure of nickel-free alloy showed a cobalt (Co) matrix with chromium (Cr) and molybdenum (Mo) regions. The HCP structure of Co has been formed due to martensitic transformation [34, 44]. None of the carbide particles are present, which corresponds healthy with the XRD peaks observed in 0-4wt.% Ni. Moreover, the compounds of cobalt (Co), chromium (Cr), molybdenum (Mo) and nickel (Ni) can be seen clearly confirming its presence (Figure 5.4) in the matrix. The lattice parameters which were detected are provided in Table 5.6. The peaks recorded in this study are similar to those from the previous studies [9, 22, 39, 45]. Scanning electron micrographs and their cor-

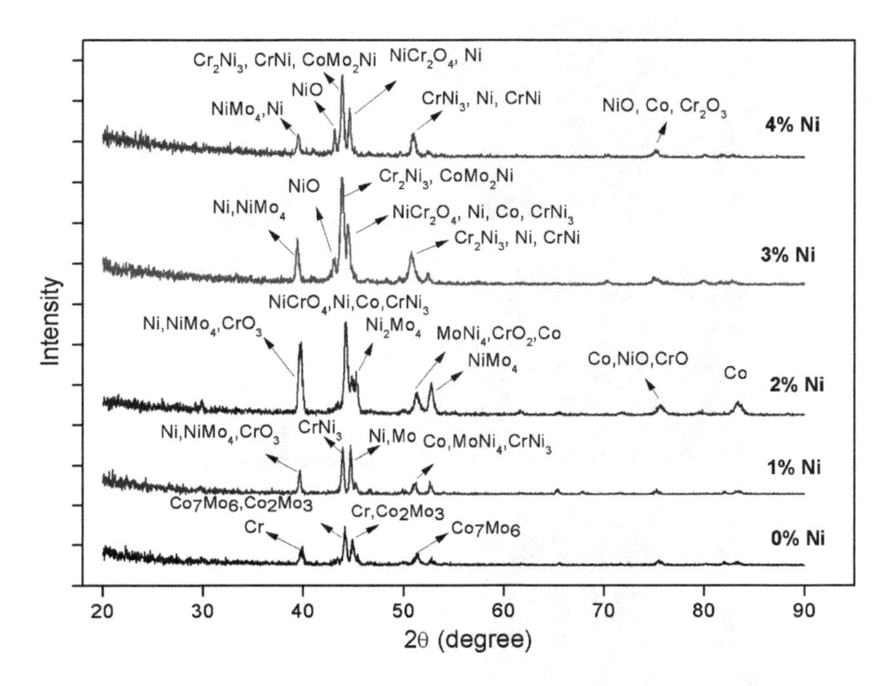

FIGURE 5.4: XRD results of the Co-30Cr-4Mo alloy with distinct nickel concentration.

responding EDX results are shown in Figure 5.5.

5.3.2 Mechanical Studies of Manufactured Material

The test results of the physical and mechanical strength of the manufactured alloys with different nickel concentration (Co-30Cr-4Mo) are listed in Table 5.7. The obtained results show that the density significantly increased with the incorporation of nickel concentration. The densities obtained from the test results are close to ASTM F75 [9].

5.3.2.1 Micro-hardness

The Vickers micro-hardness characteristics were obtained at six distinct spots and the average value was taken. Table 5.7 represents the values of micro-hardness of the manufactured alloys. It was observed that the hardness of the nickel added Co-30Cr-4Mo alloy increases linearly with the increase in nickel wt.%. It occurs due to the mixing of hard nickel particles in the base matrix. The utmost value was observed at 4wt.

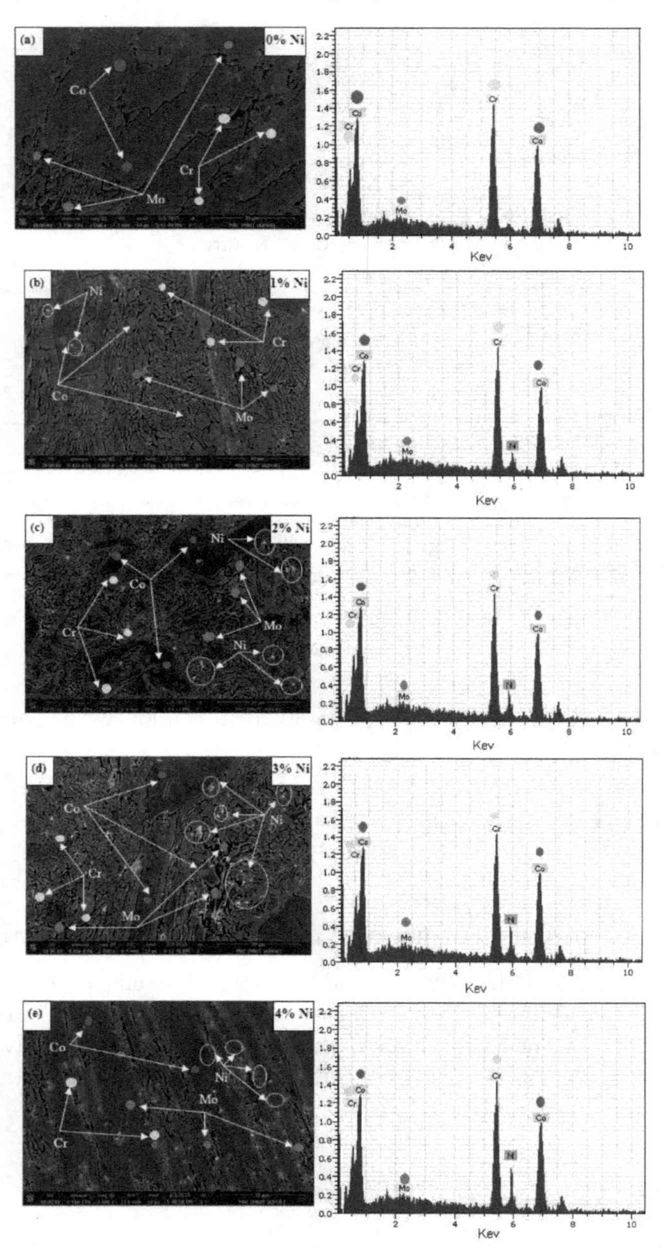

FIGURE 5.5: SEM images and related EDX results of fabricated material with different nickel content: (a) N0, (b) N1, (c) N2, (d) N3 and (e) N4. Cobalt (Co), chromium (Cr), molybdenum (Mo) and nickel (Ni) are identified. Co, Cr and Mo are shown just by white arrows; the small white spots of Ni are signified by white ovals.

TABLE 5.6: The identified phases and the lattice parameters

Phases	Crystal Structure	Lattice Parameters
Cobalt base α matrix	FCC (111)	d=2.04Å, a=b=c=3.545Å
Cr	BCC (110)	d=2.039Å, a=b=c=2.884Å
Mo	BCC (110)	d=2.225Å, a=b=c=3.147Å
CrNi3	Cubic (111)	d=2.0507Å, a=b=c=3.552Å
Co2Mo3	Tetragonal (411)	d=2.0311Å, a=b=9.229Å and c=4.827Å
Co7Mo6	Rhombohedral (116)	d=2.08Å
Co7Mo6	Rhombohedral (027)	d=1.796Å, a=b=4.762Å, and c=25.617Å
MoNi4	Tetragonal (002)	d=1.782Å, a=b=5.724Å, c=3.564Å
NiCr2O4	Cubic (220)	d=2.9407Å, a=b=c=8.318Å
NiCrO3	Rhombohedral (104)	d=2.648Å, a=b=4.925Å, c=13.504Å

TABLE 5.7: Physico-mechanical properties of manufactured orthopaedic materials

Sample	Density	Stdev	Hardness (HV)	Stdev	Compressive Strength (MPa)	Stdev
N0	8.7	0.01866	762	0.07891	1230	0.0069
N1	7.24	0.02175	590	0.48274	1080	0.0098
N2	7.8	0.01784	640	0.47539	1164	0.0073
N3	8.14	0.02156	690	0.18477	1232	0.0071
N4	8.58	0.01888	738	0.29489	1296	0.0039

5.3.2.2 Compressive strength

The compressive strength of the manufactured orthopaedic biomaterials with the Ni contents is reported in Table 5.7. It is clearly seen that from the test results, the compressive strength of manufactured materials increased as the nickel concentration increased. The utmost value was attained at 4wt.% of nickel concentration, i.e., 1296 MPa and the reason is enhancement of the hardness [20, 49].

5.3.3 Taguchi Experimental Design

Table 5.8 represents the test results for wear loss and friction coefficient of varying concentration of nickel-filled alloys and their corresponding S/N ratio values for each test. The results are converted into the S/N ratios. Each test was reiterated thrice for more accuracy and the mean value is recorded in Table 5.8. The mean S/N ratios for the entire experiment runs are 62.702 dB as computed using software MINITAB 17. Other researchers like Singh et al. [51], Sahu et al. [48], etc. also accounted for similar results in their research. Effect of input control factors for S/N ratio (see Fig. 5.6) also leads to the

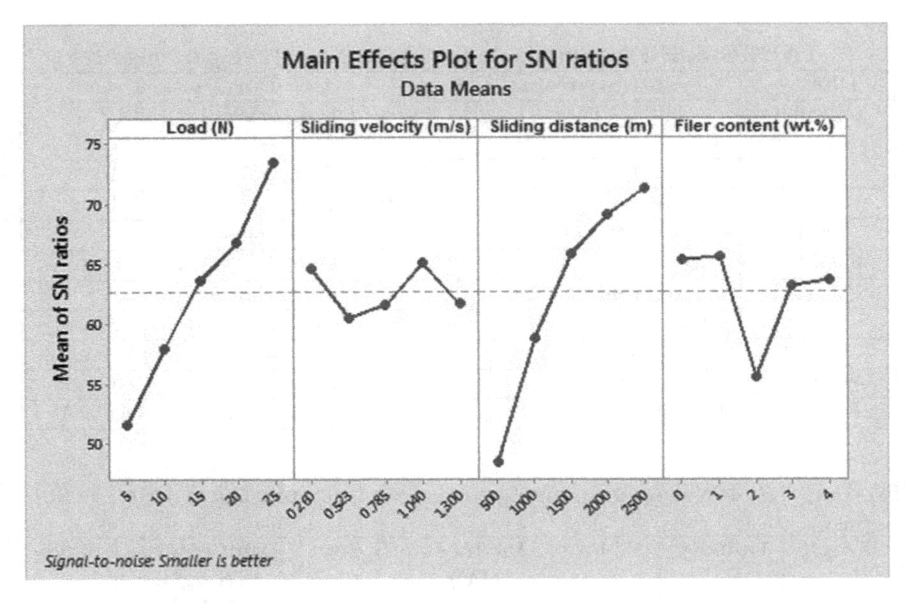

FIGURE 5.6: Effect of control factors on wear loss.

conclusion that an arbitrary factor setting $A_5B_1C_5D_2$ combination gives least value of wear characteristics.

5.4 Simulation Model for Wear Response

An artificial neural network (ANN) is a powerful biological stimulated mathematical model [25, 54], which can be employed to simplify complex problems that are hard to solve conventionally. The ANNs have been applied in various applications such as manufacturing, aerospace, robotics, electronics and telecommunications. Currently, the application of ANN modelling has increased rapidly in the field of material science and biomedical. ANN model is an intellectual approach that can resolve non-linear complex problems by learning the capability of human beings from the samples. Therefore, a number of experimental specimens were formed to train the network to assess the tribology properties of the tested biomedical alloys. A simple ANN model based on back propagation training algorithm was designed to predict the tribology properties including sliding wear and friction coefficient. The results of the ANN model shows a conformity with the experimental results. The network consists of: (1) Data processing, (2) network training, (3) testing, and (4) pre-

TABLE 5.8: Experimental design using L_{25} orthogonal array

Runs	Load (N)	Sliding Velocity (m/s)	Sliding Distance (m)	Filler Content (wt.%)	Wear (loss (mm^3)	S/N (ratio) (db)	Friction Coeff.	S/N (Ratio) (db)
1	5	0.26	500	0	0.0092000± 0.00002	40.7242	0.03	31.375
2	5	0.523	1000	1	0.0026400± 0.00001	51.5679	0.03	31.903
3	5	0.785	1500	2	0.0042800± 0.00003	47.3711	0.04	28.707
4	5	1.04	2000	3	0.0010800± 0.00002	59.3315	0.11	19.514
5	5	1.3	2500	4	0.0011520± 0.00002	58.7710	0.13	17.804
6	10	0.26	1000	2	0.0028000± 0.00003	51.0568	0.10	20.291
7	10	0.523	1500	3	0.0011267± 0.00003	58.9641	0.02	34.563
8	10	0.785	2000	4	0.0006900± 0.00001	63.2230	0.10	19.933
9	10	1.04	2500	0	0.0002200± 0.00001	73.1515	0.21	13.654
10	10	1.3	500	1	0.0068000± 0.00001	43.3498	0.13	18.020
11	15	0.26	1500	4	0.0003956± 0.00002	68.0559	0.22	13.288
12	15	0.523	2000	0	0.0002500± 0.00003	72.041	0.22	13.193
13	15	0.785	2500	1	0.0001413± 0.00003	76.9951	0.47	6.6430
14	15	1.04	500	2	0.0060000± 0.00002	44.4370	0.26	11.556
15	15	1.3	1000	3	0.0014400± 0.00001	56.8328	0.49	6.2380
16	20	0.26	2000	1	0.0001550± 0.00001	76.1934	0.25	12.200
17	20	0.523	2500	2	0.0009600± 0.00003	60.3546	0.35	9.0380
18	20	0.785	500	3	0.0019800± 0.00003	54.0667	0.38	8.4830
19	20	1.04	1000	4	0.0003750± 0.00002	68.5194	0.43	7.2740
20	20	1.3	1500	0	0.0001800± 0.00002	74.8945	0.36	8.8490
21	25	0.26	2500	3	0.0000448± 0.00002	86.9744	0.29	10.759
22	25	0.523	500	4	0.0010240± 0.00001	59.7940	0.31	10.286
23	25	0.785	1000	0	0.0004840± 0.00001	66.3031	0.41	7.7260
24	25	1.04	1500	1	0.0001013± 0.00002	79.8850	0.24	12.357
25	25	1.3	2000	2	0.0001840± 0.00001	74.7036	0.35	9.1270

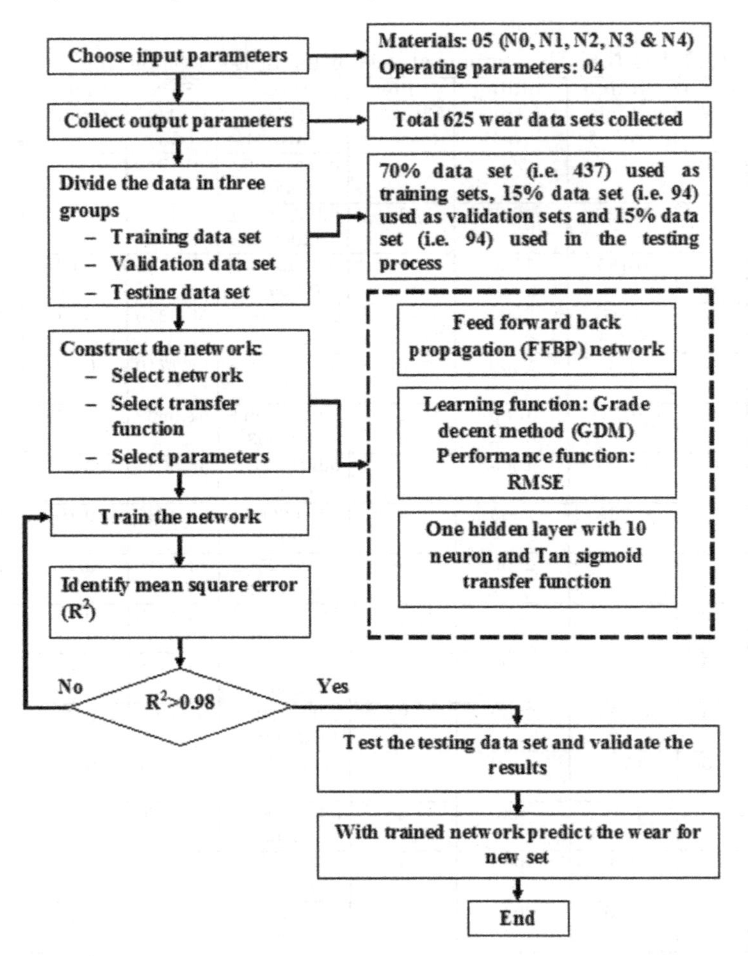

FIGURE 5.7: Flow diagram to predict the response with ANN.

dicting the response. Figure 5.7 demonstrates the flow diagram to predict the response with ANN.

5.4.1 Data Processing in ANN Model

For data processing, the input dataset which is given to test the ANN model must be numeral and the values lie in the gap of 0 and 1. The output values which are attained from neural network should also lie in the gap of 0 and 1. To make the training more efficient, some processing steps should be carried out on the network. An ANN model that demonstrates the architecture shown in Figure 5.8 is written as:

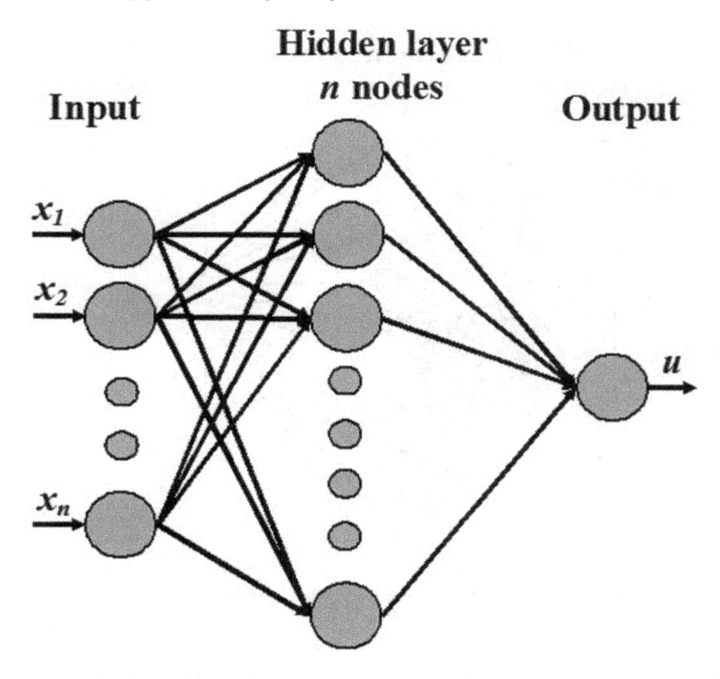

FIGURE 5.8: Architecture of a back propagation neural network.

$$G = f(x) = K_o \times tanh\,(K_i \times X + H_i) + m_o \tag{5.3}$$

Where,

G = Output (i.e., wear loss) of NN model
X = Column vector of size q (input values)
K_o = Row vector of size r from the hidden layer to the output
K_i = Matrix consist of r rows and q columns
H_i = Column vector of size q from the input to the hidden layer
m_o = Bias (scalar) from the hidden layer to the output of the NN model

Each input u_j, j=1, 2, 3.....q has lesser and greater bounds, Lm_j (least value of the jth input) and Um_j (highest value of input), respectively.

5.4.2 Network Training

The most vital part of the ANN model is the network training which is based on two important elements: the transfer function and the learning training module. The principal function of first element, i.e., transfer function is to transform the input neuron into the output neuron and these neurons

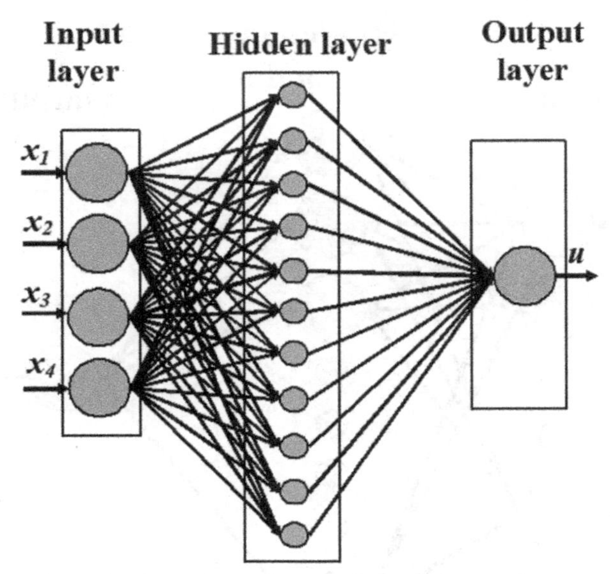

FIGURE 5.9: Schematic representation of ANN architecture applied to wear loss prediction.

may use any differentiable transfer functions (hard limit transfer function and many more) to generate their output. The second element called learning training module is a mathematical logic which improves the ANN performance and typically this is applied repeatedly over the network. In this study, back propagation (BP) learning model was utilized to design and train the neural network with multilayer perceptron (MLP) by using the MATLAB® software package.

To start the training process, first allocate the arbitrary values to all the wear factors. Once the arbitrary values are assigned, the network processes with one pair of input data. Its output was computed by utilizing the first element known as transfer function and was compared with a target output. The dissimilarity between the calculated and target values and the target values (also called prediction error) was subsequently back-propagated through the network. New wear loss values were recorded and ascribed accordingly. This cycle was replicated for whole dataset. Once the output, i.e., wear loss has been adjusted for all sets of pairs, one training epoch was successfully done. The entire cycle was reiterated, until the percentage of error was minimized. Once the network was trained, the wear test factors were set and the neural network model was used to find out the output for any combination set of input data.

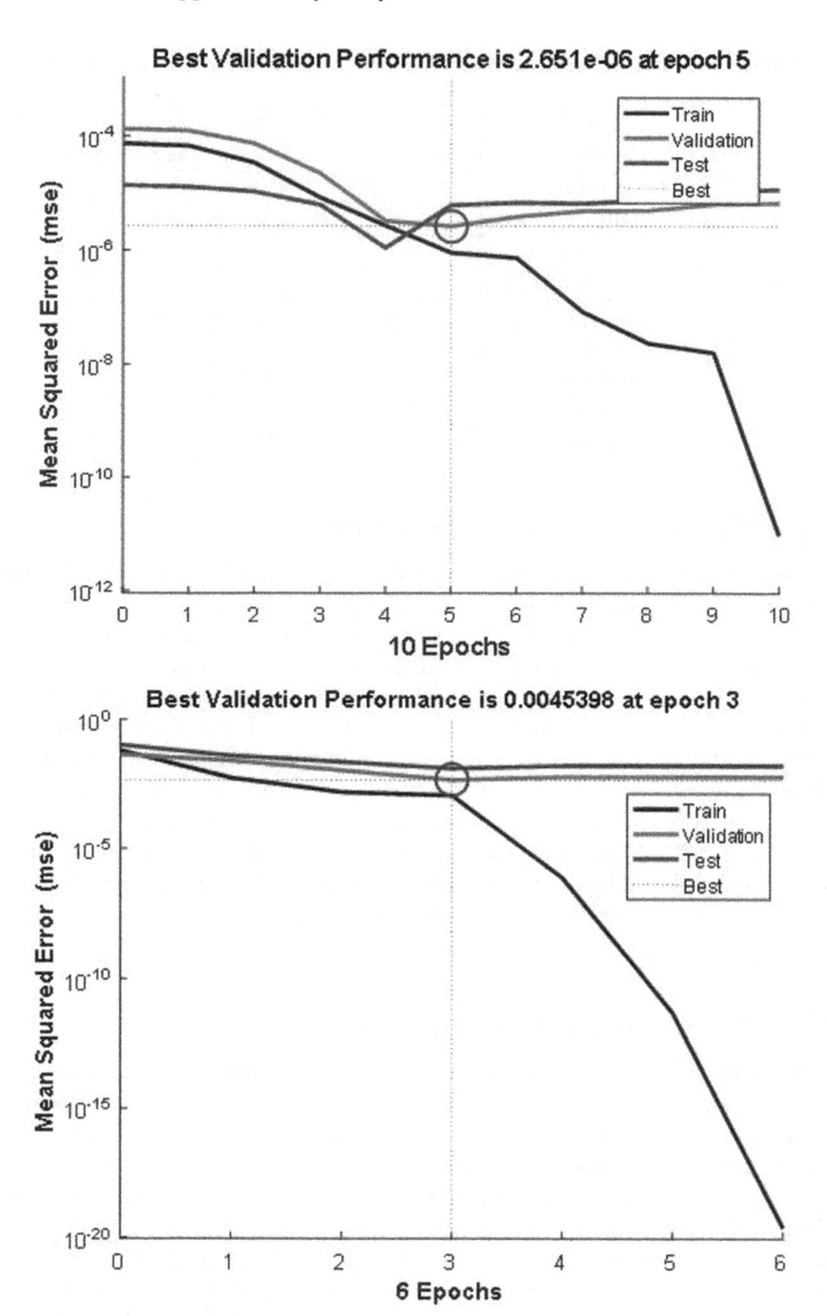

FIGURE 5.10: Neural network training, validation and testing performance system for (a) sliding wear and (b) sliding friction.

TABLE 5.9: Network architecture parameters for wear loss and friction coefficient

Type	**Values**
Architecture	Feed-forward back propagation
No. of hidden layer	10
No. of input layer	4
No. of output layer	1
Transfer function	Tan sigmoid
Training algorithm	Gradient descent learning
Learning rate	0.001
Error goal	1×10^{-4}
Epochs	10,00,000
Momentum rate	0.001

5.4.3 Neural Network Architecture

The network model to predict the tribology properties was executed based on the trained neural network as discussed in the previous step. A BP algorithm consists of a feed-forward network employed in training. This algorithm was an efficient way for training multilayer ANNs because it has no limits on the number of hidden layers. In the present work, ANN model contains 4 neurons in the input layer (see Figure 5.9), one hidden layer includes 11 neurons, and 1 output neuron has been designed to predict the wear loss for various input operating parameters. The amount of neurons present in the hidden layer was carried out by a trial and error method based on the mean square error (MSE) criterion, and it was concluded that the network contains a single hidden layer having 11 neurons that fit in the proposed NN model as shown in Figure 5.9 and it is a 4-11-1 architecture.

5.4.4 ANN Prediction and its Factor

Based on the training procedure as discussed above, the test results of sliding wear and friction were predicted by using a feed-forward BP neural network model. The architecture of NN model was constructed and includes 4 input nodes, 1 hidden layer and 1 output node called wear loss (see Table 5.9). In this study, 625 experimental data points were utilized to build a feed-forward BP network. Out of 625 datasets, 70% data (i.e., 437 datasets) considered for training, 15% data (i.e., 94 datasets) for validation and the rest 15% data for testing. To give an optimum result of NN, Tan sigmoid is considered as transfer function for the input layer and hidden layer. Whereas, purelin is considered for the output layer. Figures 5.10a and 5.10b show the best training, validation and testing performance of the neural network for sliding wear and sliding friction, with averaged MSE reaching as low as 2.651e-06 and

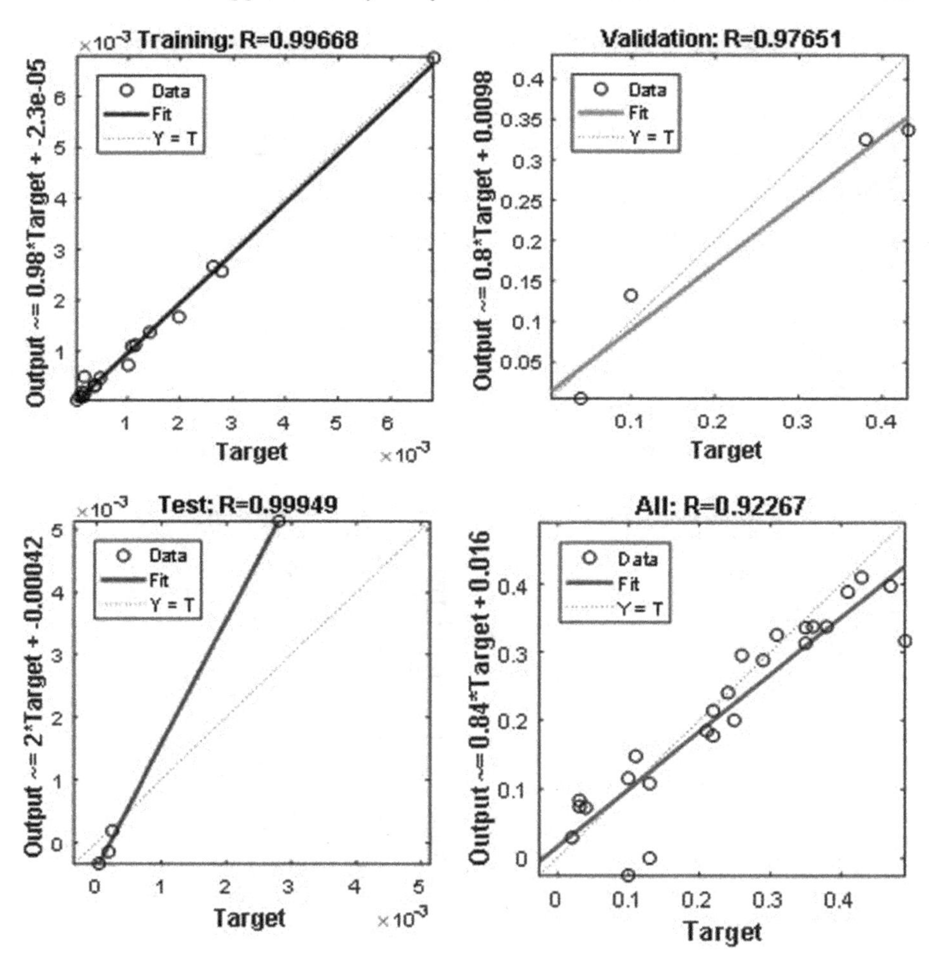

FIGURE 5.11: Coefficient of determination obtained by training a network of friction coefficient.

0.0045398. The network observes the coefficient of determination R=0.92267 and R=0.90897 for wear loss and friction coefficient, respectively, shown in Figures 5.11 and 5.12. If R=1, the network gives healthier prediction results [1]. Hence, these observed results clearly stated that the chance of using the ANN model efficiently with minimum percentage of error for the prediction of the tribology properties of the manufactured biomedical alloys. From this analysis we can say that the ANN tool is a most influencing tool to predict the output values with good accuracy. Figure 5.13 shows the comparison of the experiment results and ANN results with all the 25 runs of experiments

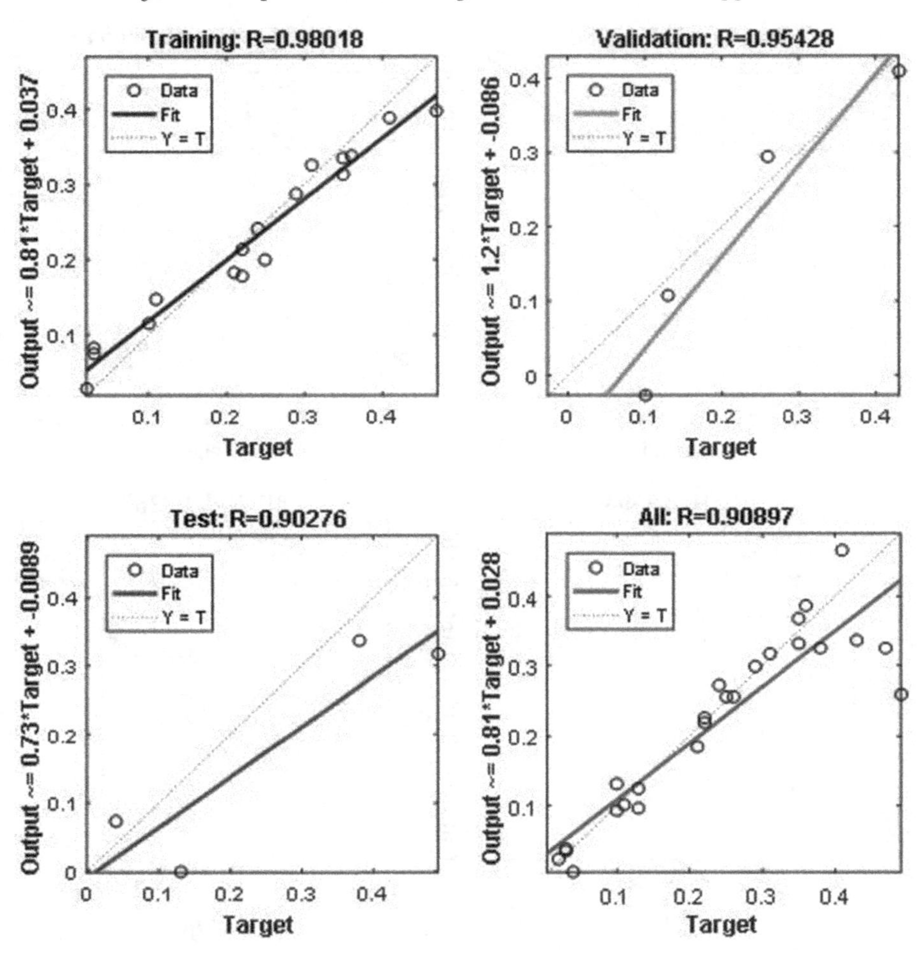

FIGURE 5.12: Coefficient of determination obtained by training a network of friction coefficient.

as listed in Table 5.10. It is clearly shown in Figure 5.13, the results obtained by experiments and the values obtained from ANN are very close to each other and the percentage errors are much fewer, which confirms the validity of the neural computation. Hence, it can be concluded that prediction by ANN tool has a potential to predict and analyze the tribological properties of orthopaedic material. Therefore, it can be recommended if it is properly trained.

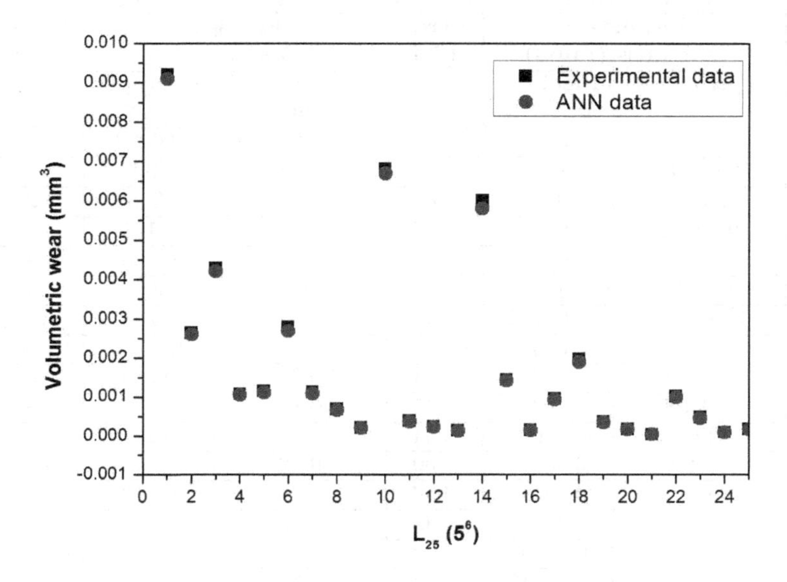

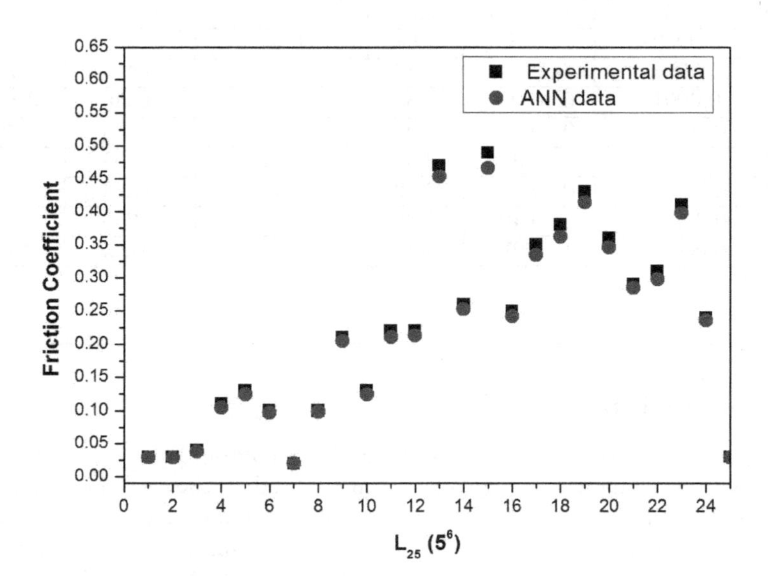

FIGURE 5.13: The comparison of measured values and ANN values for (a) Volumetric wear; (b) friction coefficient as a function of L_{25} orthogonal array.

TABLE 5.10: Experimental data and ANN test results

Test Run	Volumetric Wear Loss (mm3)		Error (%)	Friction Coefficient		Error (%)
	Experimental results	Predicted by ANN		Experimental results	Predicted by ANN	
1	0.0092	0.0091	1.08	0.03	0.0293	2.33
2	0.00264	0.00261	1.13	0.03	0.029	3.33
3	0.00428	0.00421	1.63	0.04	0.0387	3.25
4	0.00108	0.00105	2.77	0.11	0.105	4.54
5	0.00115	0.00112	2.43	0.13	0.125	3.84
6	0.0028	0.0027	3.57	0.1	0.097	3.0
7	0.001126	0.00109	2.90	0.02	0.0196	2.0
8	0.00069	0.00067	2.89	0.1	0.0984	1.60
9	0.0002	0.00021	4.54	0.21	0.205	2.38
10	0.0068	0.0067	1.47	0.13	0.125	3.84
11	0.00039	0.00038	3.94	0.22	0.211	4.09
12	0.00025	0.00024	4.00	0.22	0.213	3.18
13	0.00014	0.00014	0.92	0.47	0.453	3.61
14	0.0060	0.0058	3.33	0.26	0.253	2.69
15	0.00144	0.00142	1.38	0.49	0.466	4.89
16	0.00015	0.00015	1.93	0.25	0.242	3.20
17	0.00096	0.00094	2.08	0.35	0.335	4.28
18	0.00198	0.0019	4.04	0.38	0.362	4.73
19	0.00037	0.00036	2.66	0.43	0.414	3.72
20	0.00018	0.000174	3.33	0.36	0.346	3.88
21	0.000044	0.000043	2.27	0.29	0.285	1.72
22	0.001024	0.001	2.34	0.31	0.298	3.87
23	0.000484	0.000471	2.68	0.41	0.398	2.92
24	0.000101	0.0001	1.28	0.24	0.236	1.66
25	0.000184	0.00017	4.34	0.03	0.0288	4.0

5.5 Conclusion

This is the first study to implement the ANN computation method based on Levenberg-Marquart learning algorithm to predict the tribological properties of the manufactured orthopaedic material of nickel concentration in Co-30Cr-4Mo alloy. The use of ANN technique is to simulate the experimental results with parametric design under different operating conditions. The ANN predicted and experimental values of sliding wear and friction indicate good conformity, robustness and more accuracy. For this, 4-11-1 network structure obtained the best results for predicting the output with minimum MSE. Hence, it can be concluded that prediction by an ANN tool has a potential to predict and analyze the wear performance of orthopaedic material. Therefore, it can be recommended if it is properly trained. The ANN technique can also evade wasting material and experimental testing time and costs.

Bibliography

[1] Amit Aherwar, Amit Singh, and Amar Patnaik. Prediction of effect of tungsten filled co-30cr-4mo-1ni metal matrix biomedical composite alloy on sliding wear peculiarity using taguchi methodology and ann. *Advances in Materials and Processing Technologies*, 3(4):665–688, 2017.

[2] Amit Aherwar, Amit Singh, and Amar Patnaik. Study on mechanical and wear characterization of novel co30cr4mo biomedical alloy with added nickel under dry and wet sliding conditions using taguchi approach. *Proceedings of the Institution of Mechanical Engineers, Part L: Journal of Materials: Design and Applications*, 232(7):535–554, 2018.

[3] Amit Aherwar, Amit Singh, and Amar Patnaik. A study on mechanical behavior and wear performance of a metal–metal co–30cr biomedical alloy with different molybdenum addition and optimized using taguchi experimental design. *Journal of the Brazilian Society of Mechanical Sciences and Engineering*, 40(4):213, 2018.

[4] Amit Aherwar, Amit Singh, Amar Patnaik, and Deepak Unune. Selection of molybdenum-filled hip implant material using grey relational analysis method. In *Handbook of Research on Emergent Applications of Optimization Algorithms*, pages 675–692. IGI Global, 2018.

[5] Amit Aherwar, Amit K Singh, and Amar Patnaik. Current and future biocompatibility aspects of biomaterials for hip prosthesis. *AIMS Bioengineering*, 3(1):23–43, 2015.

[6] Amit Aherwar, Amit Kumar Singh, and Amar Patnaik. Cobalt based alloy: A better choice biomaterial for hip implants. *Trends in Biomaterials & Artificial Organs*, 30(1), 2016.

[7] Necat Altinkok and Rasit Koker. Use of artificial neural network for prediction of physical properties and tensile strengths in particle reinforced alüminum matrix composites. *Journal of Materials Science*, 40(7):1767–1770, 2005.

[8] Jorge Alvarado, Ricardo Maldonado, Jorge Marxuach, and Ruben Otero. Biomechanics of hip and knee prostheses. *Applications of Engineering Mechanics in Medicine, GED–University of Puerto Rico Mayaguez*, pages 1–20, 2003.

[9] ASTM ASTM. F75: Standard specification for cobalt-28 chromium-6 molybdenum alloy castings and casting alloy for surgical implants (uns r30075). *West Conshohocken: ASTM International*, 2007.

[10] G ASTM. 99–95a standard test method for wear testing with a pin-on-disk apparatus. *ASTM International*, 2000.

[11] S Basavarajappa, KV Arun, and J Paulo Davim. Effect of filler materials on dry sliding wear behavior of polymer matrix composites–a taguchi approach. *Journal of Minerals and Materials Characterization and Engineering*, 8(05):379, 2009.

[12] S Basavarajappa, G Chandramohan, and J Paulo Davim. Application of taguchi techniques to study dry sliding wear behaviour of metal matrix composites. *Materials & Design*, 28(4):1393–1398, 2007.

[13] S.V. Bhat. *Biomaterials* (2nd edition). Alpha Science International Lim ited, New Delhi, India, 2005.

[14] Pascal Bizot, Rfni Nizard, Sophie Lerouge, Florence Prudhommeaux, and Laurent Sedel. Ceramic/ceramic total hip arthroplasty. *Journal of Or thopaedic Science*, 5(6):622–627, 2000.

[15] Wolfram Brodner, Peter Bitzan, Vanee Meisinger, Alexandra Kaider, Florian Gottsauner-Wolf, and Rainer Kotz. Serum cobalt levels after metal-on-metal total hip arthroplasty. *JBJS*, 85(11):2168–2173, 2003.

[16] Bernard Cales. Zirconia as a sliding material: histologic, laboratory, and clinical data. *Clinical Orthopaedics and Related Research (1976-2007)*, 379:94–112, 2000.

[17] A Canakci, T Varol, and S Ozsahin. Prediction of effect of volume fraction, compact pressure and milling time on properties of al-al 2 o 3 mmcs using neural networks. *Metals and Materials International*, 19(3):519–526, 2013.

[18] Aykut Canakci, Temel Varol, and Sukru Ozsahin. Analysis of the effect of a new process control agent technique on the mechanical milling process using a neural network model: measurement and modeling. *Measurement*, 46(6):1818–1827, 2013.

[19] Fu-Kuo Chang, Jose Luis Perez, and James A Davidson. Stiffness and strength tailoring of a hip prosthesis made of advanced composite materials. *Journal of Biomedical Materials Research*, 24(7):873–899, 1990.

[20] P Choudhary, S Das, and BK Datta. Effect of ni on the wear behaviour of a zinc-aluminium alloy. *Journal of Materials Science*, 37:2103–2107, 2002.

[21] M Cruz-Romero, AL Kelly, and JP Kerry. Effects of high-pressure and heat treatments on physical and biochemical characteristics of oysters (crassostrea gigas). *Innovative Food Science & Emerging Technologies*, 8(1):30–38, 2007.

[22] ASTM F1537-11. Standard specification for wrought cobalt-28chromium-6molybdenum alloys for surgical implants (uns r31537, uns r31538, and uns r31539), 2011.

[23] Lada A Gyurova and Klaus Friedrich. Artificial neural networks for predicting sliding friction and wear properties of polyphenylene sulfide composites. *Tribology International*, 44(5):603–609, 2011.

[24] David R Haynes, Tania N Crotti, and Michael R Haywood. Corrosion of and changes in biological effects of cobalt chrome alloy and 316l stainless steel prosthetic particles with age. *Journal of Biomedical Materials Research: An Official Journal of The Society for Biomaterials and The Japanese Society for Biomaterials*, 49(2):167–175, 2000.

[25] Wei He, GH Bao, TJ Ge, AS Luyt, and XG Jian. Artificial neural networks in prediction of mechanical behavior of high performance plastic composites. In *Key Engineering Materials*, volume 501, pages 27–31. Trans Tech Publ, 2012.

[26] ASM International. *Handbook of Materials for Medical Devices*. ASM international, 2003.

[27] Farkhanda Kauser. *Corrosion of CoCrMo alloys for biomedical applications*. PhD thesis, University of Birmingham, 2007.

[28] V Kavimani and K Soorya Prakash. Tribological behaviour predictions of r-go reinforced mg composite using ann coupled taguchi approach. *Journal of Physics and Chemistry of Solids*, 110:409–419, 2017.

[29] Dongwoo Khang, Jing Lu, Chang Yao, Karen M Haberstroh, and Thomas J Webster. The role of nanometer and sub-micron surface features on vascular and bone cell adhesion on titanium. *Biomaterials*, 29(8):970–983, 2008.

[30] Wen J Ma, Andrew J Ruys, Rebecca S Mason, Phil J Martin, Avi Bendavid, Zongwen Liu, Mihail Ionescu, and Hala Zreiqat. Dlc coatings: effects of physical and chemical properties on biological response. *Biomaterials*, 28(9):1620–1628, 2007.

[31] D McMinn and J Daniel. History and modern concepts in surface replacement. *Proceedings of the Institution of Mechanical Engineers, Part H: Journal of Engineering in Medicine*, 220(2):239–251, 2006.

[32] R McMinn. Hip resurfacing (bhr) history, development and clinical results. *Midland Medical Technologies, Birmingham*, 2000.

[33] N Mirhosseini, PL Crouse, MJJ Schmidth, L Li, and D Garrod. Laser surface micro-texturing of ti–6al–4v substrates for improved cell integration. *Applied Surface Science*, 253(19):7738–7743, 2007.

[34] C Montero-Ocampo, R Juarez, and A Salinas Rodriguez. Effect of fcc-hcp phase transformation produced by isothermal aging on the corrosion resistance of a co-27cr-5mo-0.05 c alloy. *Metallurgical and Materials Transactions A*, 33(7):2229–2235, 2002.

[35] N Natarajan and RM Arunachalam. Optimization of micro-EDM with multiple performance characteristics using taguchi method and grey relational analysis. *Journal of Scientific and Industrial Research*, 70(7):500–505, 2011.

[36] Naoyuki Nomura, Mariko Abe, Atsushi Kawamura, Shigeo Fujinuma, Akihiko Chiba, Naoya Masahashi, and Shuji Hanada. Fabrication and mechanical properties of porous co–cr–mo alloy compacts without ni addition. *Materials Transactions*, 47(2):283–286, 2006.

[37] K Palanikumar, B Latha, VS Senthilkumar, and J Paulo Davim. Analysis on drilling of glass fiber–reinforced polymer (gfrp) composites using grey relational analysis. *Materials and Manufacturing Processes*, 27(3):297–305, 2012.

[38] Bhairav Patel, Gregory Favaro, Fawad Inam, Michael J Reece, Arash Angadji, William Bonfield, Jie Huang, and Mohan Edirisinghe. Cobalt-based orthopaedic alloys: relationship between forming route, microstructure and tribological performance. *Materials Science and Engineering: C*, 32(5):1222–1229, 2012.

[39] Bhairav Patel, Fawad Inam, Mike Reece, Mohan Edirisinghe, William Bonfield, Jie Huang, and Arash Angadji. A novel route for processing cobalt–chromium–molybdenum orthopaedic alloys. *Journal of The Royal Society Interface*, 7(52):1641–1645, 2010.

[40] Amar Patnaik, Alok Satapathy, and SS Mahapatra. Study on erosion response of hybrid composites using taguchi experimental design. *Journal of Engineering Materials and Technology*, 131(3):031011, 2009.

[41] S Ramakrishna, J Mayer, E Wintermantel, and Kam W Leong. Biomedical applications of polymer-composite materials: a review. *Composites Science and Technology*, 61(9):1189–1224, 2001.

[42] Jeremy J Ramsden, David M Allen, David J Stephenson, Jeffrey R Alcock, GN Peggs, G Fuller, and G Goch. The design and manufacture of biomedical surfaces. *CIRP Annals*, 56(2):687–711, 2007.

[43] FS Rashed and TS Mahmoud. Prediction of wear behaviour of a356/sicp mmcs using neural networks. *Tribology International*, 42(5):642–648, 2009.

[44] Buddy D Ratner, Allan S Hoffman, Frederick J Schoen, Jack E Lemons, Joseph Dyro, Orjan G Martinsen, Richard Kyle, Bernhard Preim, Dirk Bartz, Sverre Grimnes, et al. *Biomedical Engineering e-Mega Reference.* Academic Press, 2009.

[45] R Rosenthal, BR Cardoso, IS Bott, RPR Paranhos, and EA Carvalho. Phase characterization in as-cast f-75 co–cr–mo–c alloy. *Journal of Materials Science*, 45(15):4021–4028, 2010.

[46] Arun Rout and Alok Satapathy. Analysis of dry sliding wear behaviour of rice husk filled epoxy composites using design of experiment and ann. *Procedia Engineering*, 38:1218–1232, 2012.

[47] T Sahraoui, S Guessasma, NE Fenineche, G Montavon, and C Coddet. Friction and wear behaviour prediction of hvof coatings and electroplated hard chromium using neural computation. *Materials Letters*, 58(5):654–660, 2004.

[48] Suvendu Prasad Sahu, Alok Satapathy, Amar Patnaik, KP Sreekumar, and PV Ananthapadmanabhan. Development, characterization and erosion wear response of plasma sprayed fly ash–aluminum coatings. *Materials & Design*, 31(3):1165–1173, 2010.

[49] Temel Savaşkan and Yasin Alemdağ. Effect of nickel additions on the mechanical and sliding wear properties of al–40zn–3cu alloy. *Wear*, 268(3-4):565–570, 2010.

[50] P Senthil Kumar, K Manisekar, and R Narayanasamy. Experimental and prediction of abrasive wear behavior of sintered cu-sic composites containing graphite by using artificial neural networks. *Tribology Transactions*, 57(3):455–471, 2014.

[51] Tej Singh, Ranchan Chauhan, Amar Patnaik, Brijesh Gangil, Ramesh Nain, and Anil Kumar. Parametric study and optimization of multiwalled carbon nanotube filled friction composite materials using taguchi method. *Polymer Composites*, 39(S2):E1109–E1117, 2018.

[52] B Suresha, BN Ramesh, KM Subbaya, and G Chandramohan. Mechanical and three-body abrasive wear behavior of carbon-epoxy composite with and without graphite filler. *Journal of Composite Materials*, 44(21):2509–2519, 2010.

[53] Genichi Taguchi. Introduction to quality engineering: designing quality into products and processes. Technical report, 1986.

[54] Jack V Tu. Advantages and disadvantages of using artificial neural networks versus logistic regression for predicting medical outcomes. *Journal of Clinical Edpidemiology*, 49(11):1225–1231, 1996.

[55] H Uwe and C Gulio. Total hip arthroplasty. jrc scientific and policy reports–state of the art, challenges and prospects. *Eur Comm*, page 7, 2012.

[56] Maria Vallet-Regí. Ceramics for medical applications. *Journal of the Chemical Society, Dalton Transactions*, (2):97–108, 2001.

[57] Temel Varol, Aykut Canakci, and Sukru Ozsahin. Artificial neural network modeling to effect of reinforcement properties on the physical and mechanical properties of al2024–b4c composites produced by powder metallurgy. *Composites Part B: Engineering*, 54:224–233, 2013.

[58] A Wang, A Essner, and G Schmidig. The effects of lubricant composition on in vitro wear testing of polymeric acetabular components. *Journal of Biomedical Materials Research Part B: Applied Biomaterials: An Official Journal of The Society for Biomaterials, The Japanese Society for Biomaterials, and The Australian Society for Biomaterials and the Korean Society for Biomaterials*, 68(1):45–52, 2004.

[59] Ken-ichi Yamada, Satoshi Nakamura, Toshio Tsuchiya, and Kimihiro Yamashita. Electrical properties of polarized partially stabilized zirconia for biomaterials. In *Key Engineering Materials*, volume 216, pages 149–152. Trans Tech Publ, 2002.

[60] Chao-Lieh Yang. Optimizing the glass fiber cutting process using the taguchi methods and grey relational analysis. *New Journal of Glass and Ceramics*, 1(01):13, 2011.

[61] AF Yetim, MY Codur, and M Yazici. Using of artificial neural network for the prediction of tribological properties of plasma nitrided 316l stainless steel. *Materials Letters*, 158:170–173, 2015.

Chapter 6

Laws Energy Measure Based on Local Patterns for Texture Classification

Sonali Dash

Department of ETC, RIT, Vishakhapatnam 531162, Andhra Pradesh, India

Manas R. Senapati

Department of IT, VSSUT, Burla 768018, Odisha, India

6.1	Introduction	131
6.2	Related Work	134
	6.2.1 Mathematical Background of LBP	134
	6.2.2 LBP Minimum	136
	6.2.3 LBP Intensity	136
	6.2.4 LBP Uniform	136
	6.2.5 LBP Number	137
	6.2.6 LBP Median	137
	6.2.7 LBP Variance	138
	6.2.8 CLBP	138
	6.2.9 Sobel-LBP	139
	6.2.10 Laws' Mask	140
6.3	Local Pattern Laws' Energy Measure	140
	6.3.1 Problem Formulation	140
6.4	Implementation and Experiments	142
	6.4.1 Results of Brodatz Database	145
	6.4.2 Results of ALOT Database	146
	6.4.3 Statistical Comparison of the Methods	148
6.5	Conclusion	149
	Bibliography	150

6.1 Introduction

Texture contains the vital part of information among all the characteristics present in an image. In texture analysis, while doing the classification, the most important part is to extract texture features by utilizing texture descriptors. Many active areas of research cover texture analysis. Even though the approaches associated to the information extraction regarding texture for all the problems are similar, the subsequent usage of this information is rather different. Texture analysis covers a wide area of applications like biomedical image, remote sensing, quality control, and many more. Texture analysis techniques have been utilized for the classification, segmentation, synthesis, and shape from texture for these types of images.

A variety of quantitative texture descriptors are developed during the last decades. Texture analysis methods are categorised as statistical, geometrical, model-based and signal processing methods. Among the statistical methods, Gray Level Co-occurrence Matrix (GLCM) is the most used method [1]. The Local Binary Pattern (LBP) operator that represents together structural and statistical approaches for texture analysis is an extremely versatile operator [2, 3, 4]. Afterwards researchers have suggested numerous types of LBP variants and their many alterations and upgradations have come out in the literature such as Generalized Local Binary Patterns (GLBP), LBP variance (LBPV), Number Local Binary Pattern (NLBP), Completed LBP (CLBP), Sobel-LBP, and LBP median for the improvement of texture classification [5, 6, 7, 8, 9, 10]. Among signal processing methods, a linear transformation called Laws' texture measure [11] and wavelet representation [12] are commonly used and have been successfully implemented for many texture analyses.

A texture analysis using convolutions with different filter masks is named as Laws' masks. Experimentally it has been proved that many of these are of suitable sizes and provide very useful information for individualizing different types of textures. Laws' mask descriptor has achieved a wide acceptance in the field of medical image analysis [13, 14].

In addition, researchers have experimented on several traditional descriptors integrated with LBP and achieved better classification rates. Few examples such as LBP are integrated Discrete Wavelet Transform (DWT) for the classification of hardwood species [15]. LBP is integrated with Local Configuration Pattern (LCP), and Local Phase Quantization (LPQ) with Gaussian Image Pyramid (GIP) for the classification of hardwood species and University of Illinois Urbana Champaign (UIUC) texture database [16]. LBP is combined with GLCM for tea leaves classification [17].

Similarly, many extensions of Laws' mask approach are also available to enhance the texture classification rates. For example, Laws' mask is integrated with steerable pyramid, Laws' mask is integrated with Discrete Wavelet Transform (DWT), and Laws' mask is integrated with bilateral filter and all the suggested approaches have achieved better success rate than the original Laws' mask approach [18, 19, 20]. However, in the field of texture analysis by combining LBP and its variants with Laws' mask approach is still a challenging problem.

This work is focused on the improvement of texture classification by suggesting a new approach by integrating the LBP and its variants with Laws' mask approach for texture feature extraction. The new suggested hybrid model of feature extraction technique is named as Local Pattern Laws' Energy Measure (LPLEM). To validate the efficiency of the recommended feature extraction method, two different texture datasets (Brodatz and ALOT) are used. For comparison purposes, experiments are also carried out for the original Laws' masks approach. The experiments executed are as follows:

1. In the first experiment only level, edge, spot and ripple masks are chosen. These vectors are combined in 15 different ways to achieve convolution with LBP and variants, for example E5S5, L5E5, L5R5, etc. The energy measurement filter is comprised of a moving window operation. Some statistical descriptors with moving non-linear window operation are used as energy measurement filters, through which each pixel is substituted by comparing with its local neighborhood pixel. Three statistical filters such as mean, absolute mean and standard deviation are utilized as texture energy measures in the experiments. The mean energy measure is named as MEM, the absolute mean energy measure is named as AMEM, and the standard deviation energy measure is named as SDEM.

   The texture feature vectors collected through energy measures are accommodating a different range of values. By applying min-max normalization approach feature vectors are normalized, hence presenting it for usage for classification purposes. Subsequently the texture parameters such as energy, entropy, and absolute mean are evaluated. For 15 different convolutions, altogether we have obtained 45 features for each image. In addition, the efficiency of the LPLEM based texture feature extraction techniques have been detected on the basis of classification success rate achieved through k-NN classifier.

2. For the proposed LPLEM method, LBP and its variants are initially employed to improve robustness to noise and illumination variations of textural images. Then features are extracted through Laws' mask approach to which 11 different types of LBP algorithms are applied. Furthermore, the obtained LBP images are applied on Laws' mask approach for convolution with different sets of masks. Afterwards, the texture features are extracted by following the first step.

   For all the experiments, simple k-NN classifier is used for the verification of classification accuracy. Correspondingly, the best combination of suggested approach is determined based on classification accuracy.

   The classification results for Brodatz and ALOT textural images with k-NN classifier show that the proposed texture features selection scheme is more advantageous in comparison to the existing Laws' mask approach.

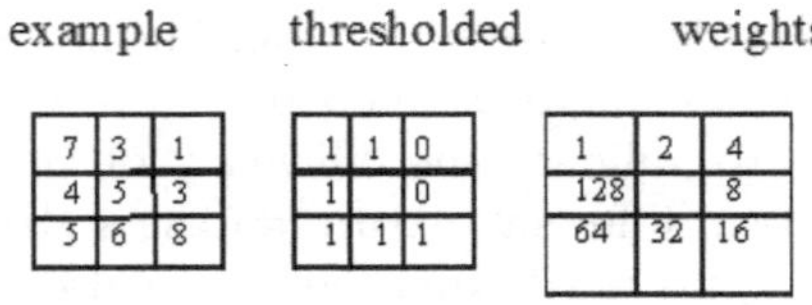

example

7	3	1
4	5	3
5	6	8

thresholded

1	1	0
1		0
1	1	1

weights

1	2	4
128		8
64	32	16

Pattern=11110011, LBP=1+2+16+32+64+128=243, C= (7+4+5+6+8+3)/6- (1+3)/2= 3.5

FIGURE 6.1: Calculation details of contrast measure and original LBP.

6.2 Related Work

The LBP is one of the most used texture descriptors in image analysis. Initially, LBP was introduced as a local descriptor. Thereafter, LBP was shown to be an interesting texture descriptor. Many extensions to the classic LBP have since then been proposed.

6.2.1 Mathematical Background of LBP

The LBP descriptor for texture analysis is described as a measure of gray-scale invariant of texture, obtained through a common description of texture in a local neighborhood. The simple idea on the background of LBP is that an image contains micro-patterns. A histogram of these micro-patterns carries the information regarding the distribution of edges and other local features in an image.

However, LBP is accompanied by an orthogonal measure of local contrast, because it is invariant to monotonic changes in gray scale. Figure 6.1 describes the procedure to derive the contrast measure (C). The average of the gray levels beneath the centre pixel is deducted from that of the gray levels above (or equal to) the center pixel. Features are taken from two-dimensional distributions and local contrast measures. The operator is called LBP/C.

Derivation: The origin of the LBP defines as follows [4]. The texture (T) is described as jointly distributing the gray levels of $P + 1$ ($P > 0$) image pixels is given below.

$$T = t(g_c, g_0, \ldots, g_{p-1}) \tag{6.1}$$

where g_c: gray value of the centre pixel of a local neighborhood.

g_p: gray values of P equally spread out pixels on a circle of radius R that create a set of neighbors which are circularly symmetric. Figure 6.2 demonstrates three neighbor sets that are circularly symmetric.

Without dropping information, g_c can be subtracted from g_p.

$$T = t(g_c, g_0 - g_c, \ldots, g_{p-1} - g_c) \tag{6.2}$$

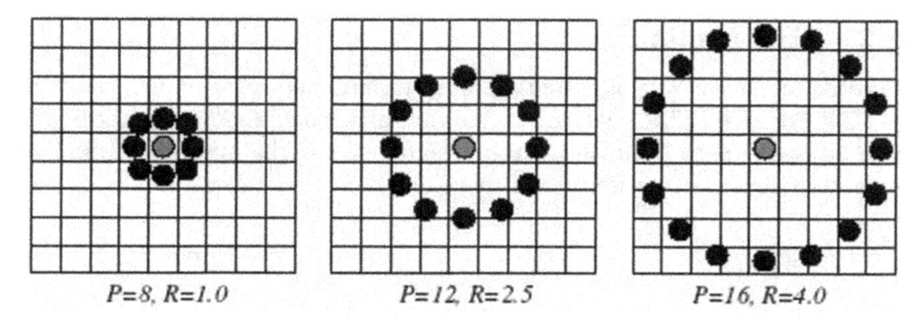

FIGURE 6.2: Examples of neighbor sets that are circularly symmetric.

With an assumption that the differences are not dependent on g_c, the following is obtained.

$$T \approx t(g_c) \times t(g_0 - g_c, \ldots, g_{p-1} - g_c) \tag{6.3}$$

Since $t(g_c)$ describes the overall luminance of an image that is unrelated to local image texture, thus can be ignored.

$$T \approx t(g_0 - g_c, \ldots, g_{p-1} - g_c) \tag{6.4}$$

Hence, LBP is described as below:

$$LBP_{P,R}(g_c) = \sum_{p=0}^{p-1} s(g_p - g_c) \times 2^p \tag{6.5}$$

where is $s(m) = \begin{cases} 1 & \text{if } m \geq 0 \\ 0 & \text{if } m < 0 \end{cases}$

There are 2^P feasible texture units relating spatial patterns in a neighborhood of P points.

Furthermore, $LBP_{P,R}$ attains invariance counter to any monotonic transformation by taking into consideration the sign of the differences in $s(g_p - g_c)$.

Merits: It has good texture discriminative property and impressive computational efficiency.

Demerits: Long histograms that are sensitive to image rotation, loss of local textural information and noise sensitivity.

The primary definition is expanded to arbitrary circular neighborhoods, and various additions to the LBP have been developed with the basic idea remaining the same. Below are described some extensions of LBP with mathematical formulation.

6.2.2 LBP Minimum

Pietikainen et al. [3] suggested an alteration named rotation invariant LBP ($LBP_{P,R}^{min}$) or ($LBP_{P,R}^{inter}$) with a view to minimize the impacts of rotation. The prime objective is to perform a circular shift to find the minimum value that the pattern chain is represented as given below

$$LBP_{P,R}^{min}(g_c) = min\{ROR(LBP_{P,R}(g_c), h)|h = 0, \ldots, P-1\} \tag{6.6}$$

where $ROR(x, h)$ executes h times of circular bit-wise operation that is right shift.

Merits: It is a very powerful approach for discriminating the local pattern contrast. Hence it performs well on local gray-scale variations.

Demerits: It does not consider the shapes of rotated binary patterns effectively.

6.2.3 LBP Intensity

Liu et al. [5] have revealed that the likelihood of a pixel, which is at center, is dependent on its neighbors. Thus, the neighbor intensity LBP ($LBP_{P,R}^{ni}$) is achieved by substituting the center of the pixel alongside the average of its neighbors described below.

$$LBP_{P,R}^{ni}(g_c) = \sum_{p=0}^{p-1} s(g_p - \mu) \times 2^p \tag{6.7}$$

where $\mu = \frac{1}{p} \times \sum_{p=0}^{p-1} g_p$

Merits: Texture classification is steadier regardless of imaging geometries of the illumination effect.

Demerits: The single intensity based descriptor (NI-LBP) does not perform well for texture classification. However, if it is combined with Radial Difference LBP (RD-LBP) and Angular Difference LBP (AD-LBP) it works well.

6.2.4 LBP Uniform

Ojala et al. [4] have noticed that a few of the LBP patterns appear more often than others, presenting like spots, curves, and flat areas. Based on these, investigation uniform patterns are described to lower the number of patterns. Uniformity measure $U(LBP_{P,R}(g_c))$ that relates to the number of spatial transition is given below.

$$U(LBP_{P,R}(g_c)) = |s(g_{p-1} - g_c) - s(g_0 - g_c)| + \sum_{p=1}^{p-1} |s(g_p - g_c) - s(g_p - g_c)| \tag{6.8}$$

In this manner, the uniform LBP ($LBP^{uni}_{P,R}$) can be achieved as:

$$LBP^{uni}_{P,R}(g_c) = \begin{cases} \sum_{p=0}^{p-1} s(g_p - g_c) & \text{if } U(LBP_{P,R}(g_c)) \leq 2 \\ P+1 & \text{Otherwise} \end{cases} \tag{6.9}$$

Merits: It preserves the frequent patterns thus reducing the number of patterns.

Demerits: This operator may not be capable of discerning textures where the dominant features occur at an extensive scale. Because combining all non-uniform patterns may eliminate some essential information.

6.2.5 LBP Number

An extension of the LBP^{uni} is recommended by Ma [6] and named as number LBP ($LBP^{num}_{P,R}$). It is achieved by the non-uniform patterns dividing into groups based on the number of "1" or "0" bits as described below.

$$LBP^{num}_{P,R}(g_c) = \begin{cases} \sum_{p=0}^{p-1} s(g_p - g_c) & \text{if } U(LBP_{P,R}(g_c)) \leq 2 \\ Num_1\{LBP_{P,R}(g_c)\} & \text{if } U(LBP_{P,R}) > 2 \text{ and} \\ & Num_1(LBP_{P,R}(g_c)) \geq \\ & Num_0\{LBP_{P,R}(g_c)\} \\ Num_0\{LBP_{P,R}(g_c)\} & \text{if } U(LBP_{P,R}) > 2 \text{ and} \\ & Num_1\{LBP_{P,R}(g_c)\} < \\ & Num_0\{LBP_{P,R}(g_c)\} \end{cases} \tag{6.10}$$

where $Num_1\{\bullet\}$ is the number of "1" and $Num_0\{\bullet\}$ is the number of "0" in the non-uniform pattern.

Merits: The texture descriptive power is improved by utilizing local structure information.

Demerits: It is limited only to describing the texture images efficiently.

6.2.6 LBP Median

Zabih [10] has recommended replacing the center of the pixel using the median of itself and the P neighbours.

$$LBP^{med}_{P,R}(g_c) = \sum_{p=0}^{p-1} s(g_p - \breve{g}) \tag{6.11}$$

where $\breve{g}$ denotes the median of the P neighbours and the central pixel. This alteration is still invariant to rotation but less sensitive to noise. Furthermore, it is invariant to illumination changes that are monotonic.

Merits: It utilizes local median in preference to using only the gray value of the centre pixel for thresholding, thus performing well for texture classification.

Demerits: Because it also codes the value of the centre pixel, it thus increases twofold the number of LBP bins.

6.2.7 LBP Variance

Previously, Ojala et al. [4] has recommended the procedure by jointly representing $LBP_{P,R}$ or $VAR_{P,R}$, where $VAR_{P,R}$ denotes the local variance. But $VAR_{P,R}$ has continuous values, hence quantization is necessary. Guo et al. [7] have incorporated complementary information of the local contrast in a new proposal named local binary pattern variance ($LBPV_{P,R}$). A rotation invariant measure of the local variance can be described as below.

$$VAR_{P,R}(g_c) = \frac{1}{p} \times (g_p - u)^2 \tag{6.12}$$

where $\{g_p | p = 0, \ldots, P-1\}$ are the g_c neighbours and $u = \frac{1}{p} \times \sum_{p=0}^{p-1} g_p$
Histogram of the LBP generally does not incorporate the information regarding the variance $VAR_{P,R}$.

The $LBPV_{P,R}$ approach provides a solution for that as given below.

$$LBPV_{P,R}(k) = \sum_{i=1}^{N}\sum_{j=1}^{M} w(LBP_{P,R}(i,j),k), k \in [0, K] \tag{6.13}$$

$$w(LBP_{P,R}(i,j),k), k = \begin{cases} VAR_{P,R}(i,j) & LBP_{P,R}(i,j) = k \\ 0 & \text{Otherwise} \end{cases} \tag{6.14}$$

The variance $VAR_{P,R}$ is utilized as an adaptive weight to regulate the supplement of the LBP code for histogram computation.

Merits: It incorporates local contrast information by using the variance as a locally adaptive weight to regulate the supplement of each LBP code.

Demerits: It escapes the quantization pre-training utilized.

6.2.8 CLBP

Guo et al. [8] have suggested the Completed LBP (CLBP) approach for the improvement of the capability of the LBP approach. The image local difference is divided into two complementary components: the sign component and the magnitude component. Consider the following:

$$s_p = s(i_r - i_q\ ), m_p = |i_r - i_{cq}| \tag{6.15}$$

Afterwards the s_p is utilized to construct the CLBP-Sign ($CLBP\_S$), while the m_p is utilized to construct the CLBP-Magnitude ($CLBP_M$). The $CLBP_S$ and $CLBP_M$ are defined below.

$$CLBP\_S_{P,R} = \sum_{p=0}^{p-1} 2^p s(i_r - i_q), s = \begin{cases} 1 & i_r \geq i_q \\ 0 & i_r < i_q \end{cases} \tag{6.16}$$

$$CLBP\_M_{P,R} = \textstyle\sum_{p=0}^{p-1} 2^p t(m_p, c)$$

$$t(m_p, c) = \begin{cases} 1 & |i_r - i_q| \geq c \\ 0 & |i_r - i_q| < c \end{cases} \tag{6.17}$$

where i_q defines the gray measure of the center pixel and i_r ($p = 0, \ldots, P-1$) defines the gray measure of the neighbor pixel on a circle of radius R and P is the number of the neighbors and c represents the mean value of m_p of the entire image.

The $CLBP\_S$ is equivalent to LBP, while the $CLBP\_M$ calculates the local variance of magnitude.

Merits: It includes both sign and magnitude information of the local region. It combines multiple LBP type features through a joint histogram for texture classification.

Demerits: More image local structural information is preserved by the sign component rather than the magnitude component.

6.2.9 Sobel-LBP

Zhao et al. [9] have suggested the Sobel-LBP approach. This approach is an easy and effective implementation, through which the computational work of LBP feature extraction process is increased. It consists of two 3 × 3 kernels (horizontal kernel S_x and vertical kernel S_y) that are convolved with the original image F for the calculation of gradient approximations:

$$F^x = S_x \times F = \begin{bmatrix} 1 & 0 & -1 \\ 2 & 0 & -2 \\ 1 & 0 & -1 \end{bmatrix} \times F, F^y = S_y \times F = \begin{bmatrix} 1 & 2 & 1 \\ 0 & 0 & 0 \\ -1 & -2 & -1 \end{bmatrix} \times F \tag{6.18}$$

F^x and F^y denote the horizontal and vertical filtered results, respectively. Normally F^x and F^y are combined to deliver the gradient magnitude $\sqrt{(F^x)^2 + (F^y)^2}$. At this point, the Sobel-LBP approach is described as the concatenation of LBP operations on I^x and I^y.

$$Sobel\_LBP_{P,R} = \{Sobel\_LBP^x_{P,R}, Sobel\_LBP^y_{P,R}\} \tag{6.19}$$

where

$$Sobel\_LBP^x_{P,R} = \sum_{p=0}^{p-1} s(F^x_{P,R} - F^x_c)2^p \tag{6.20}$$

$$Sobel\_LBP^y_{P,R} = \sum_{p=0}^{p-1} s(F^x_{P,R} - F^y_c)2^p \tag{6.21}$$

Merits: It identifies all significant edge pixels in the image.

Demerits: It does not consider rotation invariant, illumination changes or any other improvement in the image except edge.

6.2.10 Laws' Mask

In 1980, Kenneth Ivan Laws suggested an approach for the extraction of texture features that describe texture properties such as roughness, density, regularity, uniformity, etc., by utilizing a set of filters [11]. Originally he has recommended 1×3 vectors such as $L3 = [1, 2, 1]$ for averaging, $E3 = [-1, 0, 1]$ for edges and $S3 = [-1, 2, -1]$ for spots. One-dimensional vectors are convolved with each other or in turn transposes with themselves that create 1×5 vectors with a mnemonics Level, Edge, Spot, Wave, and Ripple as given below.

$Level \rightarrow L5 = [1, 4, 6, 4, 1] = L3 \times L3$
$Spot \rightarrow S5 = [-1, 0, 2, 0, -1] = E3 \times E3 = L3 \times S3$
$Ripple \rightarrow R5 = [1, -4, 6, -4, 1] = S3 \times S3$
$Edge \rightarrow E5 = [-1, -2, 0, 2, 1] = L3 \times E3$
$Wave \rightarrow W5 = [-1, 2, 0, -2, 1] = E3 \times S3$

where $*$ symbolizes the convolution operation. Different two-dimensional convolution masks of 25 numbers can be generated by the convolution of each vertical vector with a horizontal one. These two-dimensional masks are known as Laws' masks that are convolved with texture image and subsequently energy metrics of the texture image are obtained.

Merits: The set of filters used in Laws' mask perform efficiently in extracting the texture parameters.

Demerits: It considers neither the rotation variations nor the illumination variations of the images.

6.3 Local Pattern Laws' Energy Measure

In this approach, LBP and the extensions of the LBP are combined with Laws mask thus named as Local Pattern Laws' Energy Measure (LPLEM).

6.3.1 Problem Formulation

As outlined in the introduction regarding the proposed approach, we have utilized the LPLEM technique to take out features from textural images for classification. The LPLEM descriptor employs LBP variant operators to generate an LBP variant code for each pixel of an image which results in local pattern variant image, then Laws' masks are applied on it for convolution. Fifteen Laws' masks (5×5) are used for convolution. Then the outputs obtained from the convolution process are applied to texture energy measurement filters to obtain feature vectors for the analysis of the texture property. After min-max normalization, the classification is done by utilizing the k-NN classifier. The procedure of the suggested approach is represented in Figure 6.3.

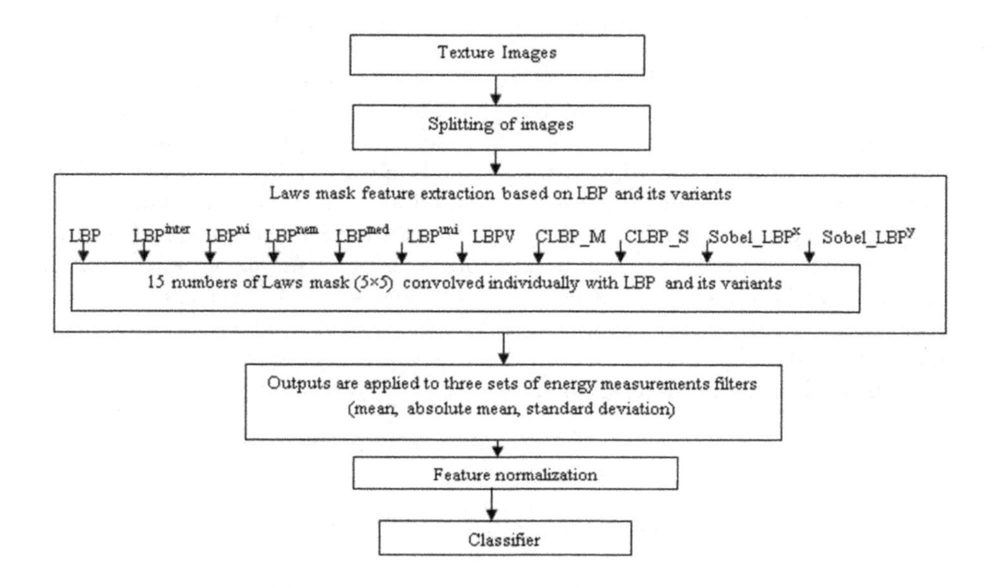

FIGURE 6.3: Block diagram of suggested approach.

TABLE 6.1: The LPLEM feature extraction techniques

Abbreviation	**Technique**
LBPLEM	Local binary pattern based Laws' energy measure
LBP^{inter}LEM	Inter local binary pattern based Laws' energy measure
LBP^{ni}LEM	Intensity local binary pattern based Laws' energy measure
LBP^{num}LEM	Number local binary pattern based Laws' energy measure
LBP^{med}LEM	Median local binary pattern based Laws' energy measure
LBP^{uni}LEM	Uniform local binary pattern based Laws' energy measure
LBPVLEM	Variance local binary pattern based Laws' energy measure
CLBP_MLEM	Complete magnitude local binary pattern based Laws' energy measure
CLBP_SLEM	Complete sign local binary pattern based Laws' energy measure
Sobel_LBP^xLEM	Sobel horizontal local binary pattern based Laws' energy measure
Sobel_LBP^yLEM	Sobel vertical local binary pattern based Laws' energy measure

Thus, different LPLEM techniques are proposed according to the combinations of different LBP variants with Laws' mask descriptor and they are listed in Table 6.1.

In this experiment only level, edge, spot and ripple are chosen. These vectors are combined in 15 different ways to achieve convolution with LBP and variants, for example E5S5, L5E5, L5R5, etc. The energy measurement filter is comprised of a moving window operation. Some statistical descriptors with moving non-linear window operation are used as energy measurement filters, through which each pixel is substituted by comparing with its local neighborhood pixel. Three statistical descriptors such as mean, absolute mean and standard deviation are utilized as texture energy measures in the experiments. The mean energy measure is named as MEM, the absolute mean energy measure is named as AMEM, and the standard deviation energy measure is named as SDEM. The filters are described below.

$$MEM = \frac{\sum_L \text{Neighbouring pixels}}{L} \tag{6.22}$$

$$AMEM = \frac{\sum_L abs(\text{Neighbouring pixels})}{L} \tag{6.23}$$

$$SDEM = \sqrt{\frac{\sum_L (\text{Neighbouring pixels} - \text{mean})^2}{L}} \tag{6.24}$$

where window size is represented by L. The texture features collected through energy measures are accommodating a different range of values. By applying min-max normalization approach, feature vectors are normalized, hence presenting it for usage of classification purposes. Subsequently the texture parameters such as energy, entropy, and absolute mean are evaluated. For 15 different convolutions, altogether we have obtained 45 features for each image. In addition, the efficiency of the LPLEM based texture feature extraction techniques have been detected on the basis of classification success rate achieved through k-NN classifier. Correspondingly, the best combination of suggested approach is determined based on classification accuracy.

6.4 Implementation and Experiments

The prime objective of this work is to suggest LPLEM based feature extraction techniques and to study their efficiency in terms of classification with an aim to get improved classification rate compared to the original Laws' masks approach. To verify the efficiency of the recommended LPLEM techniques, experiments are conducted with Brodatz and ALOT databases. For comparison purposes, experiments are also carried out for the original Laws' mask, LBP and LBP variants by using the same texture databases.

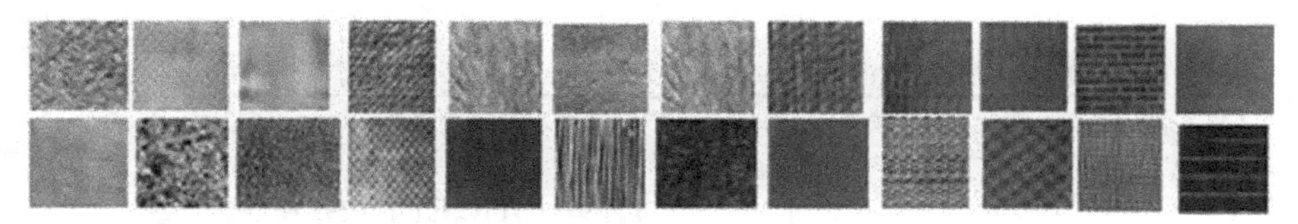

FIGURE 6.4: 24 classes of Brodatz database.

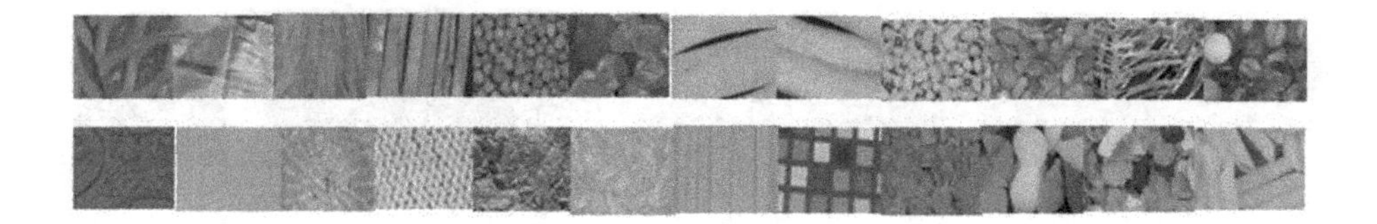

FIGURE 6.5: 24 classes of ALOT database.

The Brodatz dataset is a very old and challenging platform for classification, presentation and analysis due to the exciting multiplicity and perceptual similarity of some textures. The Brodatz dataset is downloaded from the site www.ux.uis.no/~tranden/brodatz. In this database, some of the textures belong to the same class but at different scales, while some others are very inhomogeneous that a human observer may not be able to group their samples appropriately. Considering this fact, 24 images, each of size 640 × 640 pixels are chosen arbitrarily and are presented in Figure 6.4. Each is partitioned into 25 non-overlapping patches of size 128 × 128 pixels. Thus, 600 patches are obtained from all 24 images. From 25 patches of each image, 12 patches and 13 patches are selected for training and testing, respectively.

ALOT database contains both color texture and gray texture images and is available on the website of ALOT database. The database is downloaded from the Amsterdam Library of Textures (ALOT) link: http://aloi.science.uva.nl/public_alot/.

The images are available in three types of resolutions: full, half and quarter. We have collected grey half-resolution database where each image is of size 768 × 512 pixels. This database consists of 250 texture classes. In each class, 100 texture images with varying viewpoint and illumination positions, and one additional illumination spectrum are available. Moreover, the significant height variation of some textures of materials causes large and variable shadows, which makes the classification even more difficult. Out of 250 texture classes, 24 different classes have been chosen and from each class randomly 25 texture images are selected with varying illuminations and rotations. Some of them are shown in Figure 6.5. In total, we have selected 600 (i.e., 25 × 24) images of size 768 × 512 out of which 288 images are used for training, and 312 images are used for testing.

TABLE 6.2: Performance results of the existing Laws' masks approach

Datasets	Features Extracted	Classification Rate (%)		
		MEM	AMEM	SDEM
Brodatz Masks (15)	45	93.27	86.22	89.42
ALOT Masks (15)	45	92.51	88.27	89.03

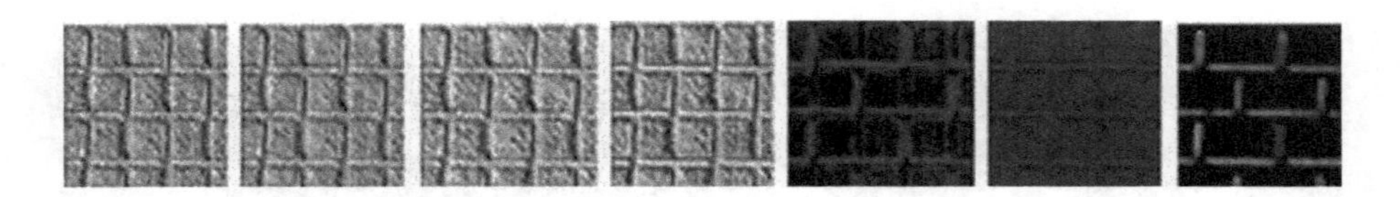

FIGURE 6.6: Few examples of images obtained from LBP and its variants.

Firstly, the original Laws' masks technique is utilized for feature extraction. By utilizing four masks: Level (L), Edge (E), Spot (S) and Ripple (R), fifteen different masks are produced. The convolution is carried out for the training and testing images using these fifteen masks. Subsequently, the outputs obtained from the convolution process are applied to the different energy measures named as MEM, AMEM and SDEM of size 15 $\times$ 15, and after normalization the texture parameters are calculated. For each image, 45 features are obtained. Table 6.2 represents the success rates of the conventional Laws' mask technique. It is observed that the MEM has delivered the maximum classification rate of 93.27% on Brodatz dataset and 92.51% on ALOT dataset. The AMEM has delivered the classification rate of 86.22% on Brodatz dataset and 88.27% on ALOT dataset. The SDEM has delivered the classification rate of 89.42% on Brodatz dataset and 89.03% on ALOT dataset.

Secondly, conventional LBP and its variants are used for feature extraction in a very simple manner. The textural images are passed through LBP and its extensions. Some of the LBP images obtained are shown in Figure 6.6.

Three statistical parameters, entropy, energy, and absolute mean, are calculated by using the LBP transformed images. We have not used the histograms of LBP. For Brodatz database, the highest classification rate of 55.77% is provided by uniform LBP. The second highest classification rate of 55.13% is provided by LBP and number LBP. For ALOT database, the number LBP has delivered the best classification rate of 56.14%. Second highest classification rate of 54.81% is delivered by uniform LBP. For LBP and its variants, sampling point as 8 and radius as 1 have been taken into account. For both databases, the classification results of the traditional LBP and variants are presented in Table 6.3.

The performance results of the original Laws' mask approach are shown in Figure 6.7 for two different datasets. The performance evaluations of the

TABLE 6.3: Performance rates of the existing LBP and its variants

Different LBP Descriptors	Classification Accuracy (%) on Brodatz Dataset	Classification Accuracy (%) on ALOT Dataset
LBP	55.13	37.50
LBP^{inter}	50.96	32.69
LBP^{med}	47.12	30.13
LBP^{ni}	52.88	28.53
LBP^{num}	55.13	**56.41**
LBP^{uni}	**55.77**	54.81
LBPV	30.13	23.08
CLBP_M	28.53	16.99
CLBP_S	19.23	20.83
Sobel_LBP^{x}	53.21	22.76
Sobel_LBP^{y}	42.31	26.28

original LBP and its extensions for both datasets are presented in Figure 6.8.

Thirdly, features are extracted from the proposed LPLEM technique. The textural images are passed through LBP and its extensions to achieve the LBP transformed images. Afterwards in the convolution process, with Laws' mask approach these transformed images are utilized. The outputs obtained from the convolution process are applied to the three energy measurement filters such as MEM, AMEM, and SDEM, and texture features are computed. The experimental results of classification accuracies by using different proposed approaches for both Brodatz and ALOT databases are listed in Table 6.4. From the results, it has been observed that out of three energy measurement filters, the SDEM has always produced highest classification accuracy among the three. Hence, the results discussed below are only for the SDEM.

6.4.1 Results of Brodatz Database

It is noted that among all the recommended methods LBP^{ni}LEM shows an outstanding performance. The classification accuracy rate has reached at 95.63%. Both LBP^{inter}LEM and LBP^{med}LEM methods accomplish the second best classification rate of 92.55%. Classification accuracy of 92.55% is delivered by both methods. For LBP^{num}LEM classification accuracy obtained is 91.46%. LBPLEM method has provided classification accuracy of 90.22%. All the better classification accuracies are obtained for the SDEM. In the proposed technique, the MEM and AMEM deliver lower classification accuracies. Figure 6.9 shows the graphical presentation of the proposed technique on Brodatz database. All the rest of the proposed techniques give lower classification rates than the original Laws' mask method.

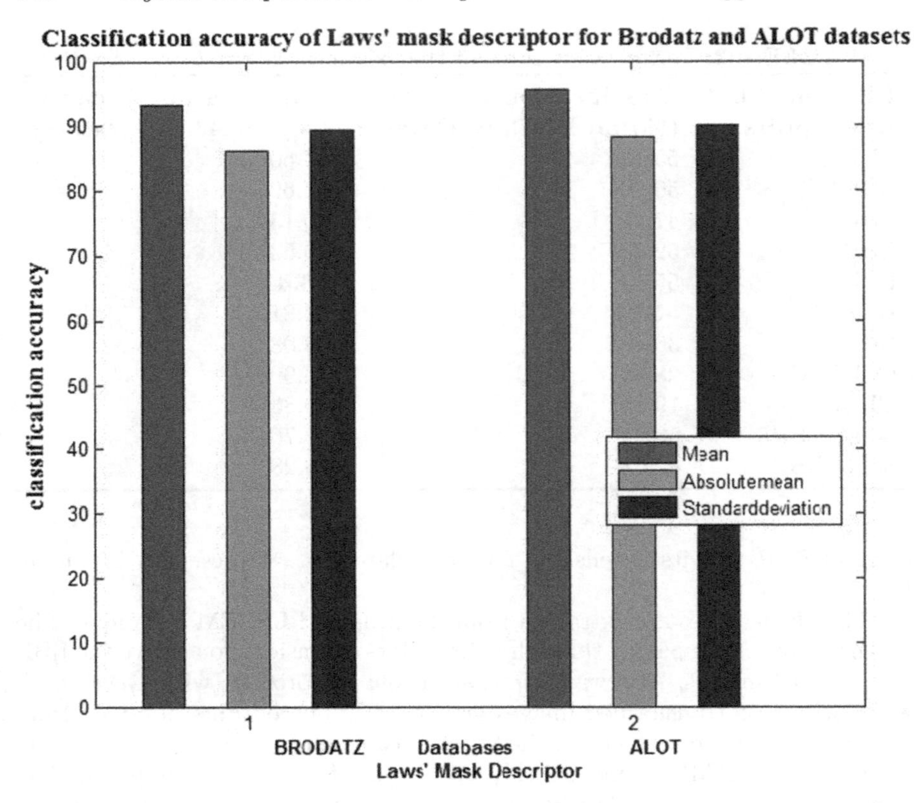

FIGURE 6.7: Performance of Laws' mask descriptor.

6.4.2 Results of ALOT Database

For ALOT database, the suggested approach $LBP^{ni}LEM$ delivers highest classification accuracy of 94.77%. The $LBP^{num}LEM$ approach has delivered second best classification rate of 93.01%. The third highest classification rate of 92.66% is attained for $LBP^{inter}LEM$ approach. LBPLEM and $LBP^{med}LEM$ also provide better classification accuracy of 90.22% and 91.19%, respectively. All the better classification rates are attained for the SDEM. For the MEM and the AMEM lower classification accuracies are achieved. Figure 6.10 depicts the graphical presentation of the proposed technique on ALOT database. All the rest of the proposed techniques deliver lower classification accuracies than the original Laws' mask approach.

It is noticed from Table 6.4 that among the three measurement filters the best classification rates are always obtained for the SDEM. The reason is that the standard deviation is a measure to quantify the amount of variation. Thus, the above analysis suggests that among all the LPLEM approaches the

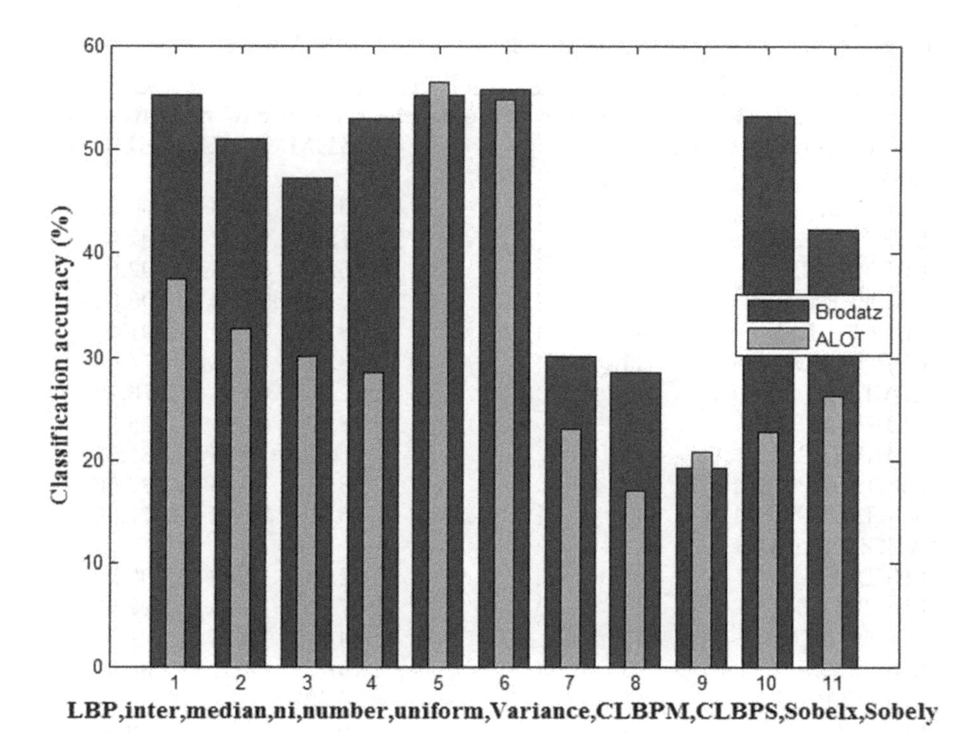

FIGURE 6.8: Performance of LBP and extensions.

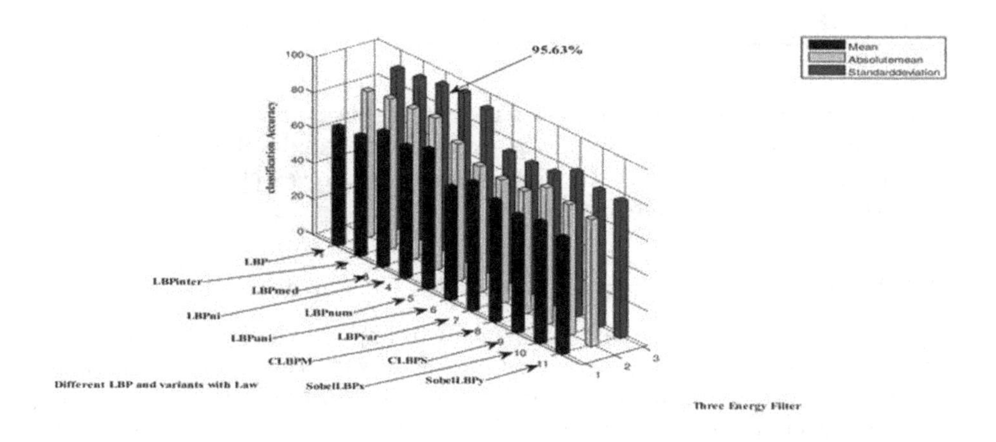

FIGURE 6.9: Accuracy of LPEM approaches on Brodatz.

TABLE 6.4: Classification results for Local Pattern Laws' Energy Measure (LPEM) for 15 different masks

Different LPLEM Techniques Brodatz Dataset	Number of Features	Classification Rates (%)		
		MEM	AMEM	SDEM
LBPLEM	45	66.67	82.69	90.22
$LBP^{inter}LEM$	45	67.95	84.62	92.55
$LBP^{med}LEM$	45	76.6	85.26	92.55
$LBP^{ni}LEM$	45	74.68	86.22	95.63
$LBP^{num}LEM$	45	78.85	77.88	91.46
$LBP^{uni}LEM$	45	63.78	71.15	74.04
LBVLEM	45	72.76	69.86	73.72
CLBP_MLEM	45	68.59	69.55	75
CLBP_SLEM	45	66.35	77.56	81.73
$Sobel\_LBP^{x}LEM$	45	68.91	74.36	77.56
$Sobel\_LBP^{y}LEM$	45	66.35	72.12	77.56
ALOT Dataset				
LBPLEM	45	72.44	74.36	90.22
$LBP^{inter}LEM$	45	67.95	78.53	92.66
$LBP^{med}LEM$	45	64.74	72.76	91.19
$LBP^{ni}LEM$	45	75.12	85.32	94.77
$LBP^{num}LEM$	45	75.12	80.71	93.01
$LBP^{uni}LEM$	45	60	76.6	79.49
LBVLEM	45	72.14	63.46	73.17
CLBP_MLEM	45	63.78	60.9	70.38
CLBP_SLEM	45	65.71	76.28	81.73
$Sobel\_LBP^{x}LEM$	45	52.56	56.41	71.32
$Sobel\_LBP^{y}LEM$	45	46.47	62.82	72.66

$LBP^{ni}LEM$ gives the best success rate for both datasets than the conventional Laws' mask method. In addition, the experimental results illustrate that all the suggested LPLEM approaches have achieved improved classification rates than the original LBP and its variants.

6.4.3 Statistical Comparison of the Methods

As the suggested approaches are focused to improve the classification rate of Law's mask, thus the comparisons are made with the results of original Laws' method taken from the original report [11]. All the statistical comparisons are given in Table 6.5. The classification rates mentioned in the table are directly quoted from the original papers of the corresponding authors. All the comparisons presented in the table are only for sampling point as 8 ($P = 8$)

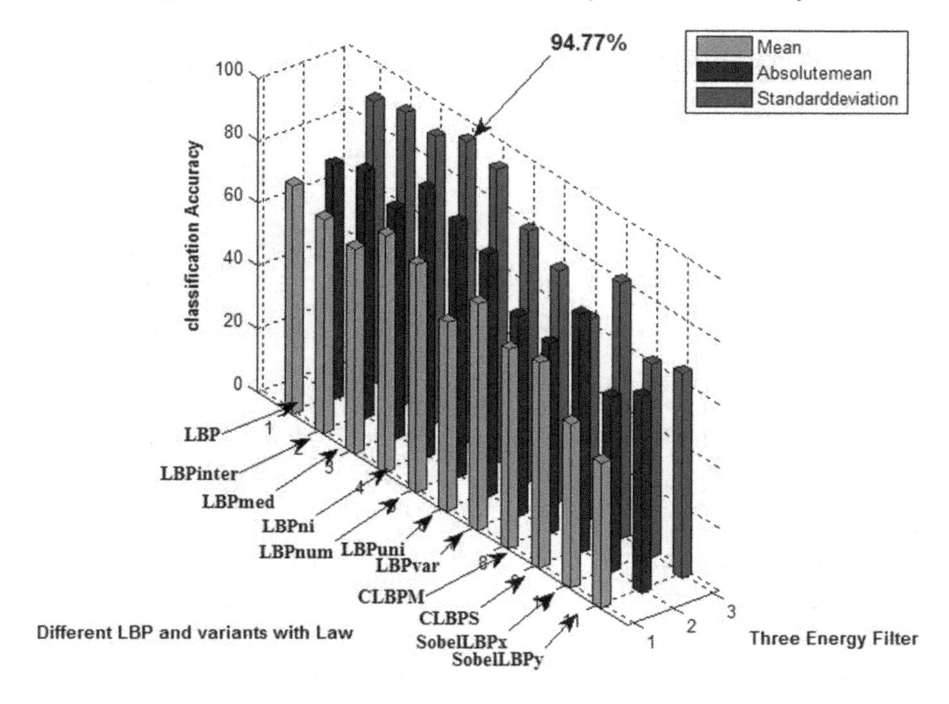

FIGURE 6.10: Accuracy of LPEM approaches for ALOT.

and radius as 1 ($R = 1$) because the experiments conducted for the suggested approaches are of the same sampling point and radius.

6.5 Conclusion

This chapter extends the existing LBP with its variants and Laws' mask approach to present a new descriptor. Laws has recommended five labelled vectors that are combined to obtain two-dimensional convolution kernels. When the texture images are convolved with these masks, individual structural components of the image are extracted. In this chapter, the LPLEM techniques for texture features extraction are suggested to enhance the classification rates over the conventional Laws' mask approach. LBP and its variants are initially employed to improve robustness to noise and illumination variations of textural images. Then features are extracted through the Laws' mask approach to which 11 different types of LBP algorithms are applied. The classifica-

TABLE 6.5: Statistical comparison of the methods

Authors	**Technique**	**Database**	**Classification Accuracy (%)**
K. I. Laws [11]	Texture energy measure	Brodatz	94
Pietikainen et al. [3]	LBP minimum	Brodatz	48.1
Ojala et al. [4]	LBP uniform	Brodatz	88.2
Liu et al. [5]	NI-LBP	Brodatz	76.7
Liu et al. [5]	RD-LBP	Brodatz	80.00
Liu et al. [5]	RD-LBP/CI (Central pixels intensity)-LBP	Brodatz	92.2
Liu et al. [5]	NI-LBP/CI-LBP	Brodatz	82.3
Liu et al. [5]	NI-LBP/RD-LBP	Brodatz	82.2
Liu et al. [5]	NI-LBP/RD-LBP/CI-LBP	Brodatz	89.4
Y. Ma [6]	LBP number (NLBP)	Brodatz	59.30
Guo et al. [7]	LBP/VAR	Outex (TC10, TC12 "t184", TC12 "horizon")	96.66, 79.25, 77.98
Guo et al. [8]	CLBP_S	Outex (TC10, TC12)	84,81, 71.3
Guo et al. [8]	CLBP_M	Outex (TC10, TC12)	81.74, 67.93
Proposed method	$LBP^{ni}LEM$	Brodatz, ALOT	**95.63, 94.77**
Proposed method	$LBP^{inter}LEM$	Brodatz, ALOT	**92.55, 92.66**
Proposed method	$LBP^{med}LEM$	Brodatz, ALOT	**92.55, 91.19**

tion results for Brodatz and ALOT textural images with k-NN classifier show that the proposed texture features selection scheme is more advantageous in comparison to the existing Laws' mask technique. The maximum increment in success rate is 6.21% on Brodatz, and 5.74% on ALOT are attained by employing the proposed techniques.

The accuracy rates of all the approaches are achieved with k-NN classifier, because the focus of the suggested method is on the effectiveness of the descriptor. In the future, utilizing more advanced SVM (support vector machine) classifier with a different database like facial database, the proposed method may improve the performance significantly.

Bibliography

[1] Haralick, R. M., K. Shanmugan, and I. Dinstein. 1973. Textural features for image classification. *IEEE Transactions on Systems Man and Cybernetics* 3: 610–612.

[2] Ojala, T., M. Pietikainen, and D. Harwood. 1996. A comparative study of texture measures with classification based on feature distributions. *Pattern Recognition* 29(1): 51–59.

[3] Pietikainen, M., T. Ojala, and Z. Xu. 2000. Rotation-invariant texture classification using feature distributions. *Pattern Recognition* 33(1): 43–52.

[4] Ojala, T., M. Pietikainen, and T. Maenpaa. 2002. Multi resolution gray-scale and rotation invariant texture classification with local binary patterns. *IEEE Transactions on Pattern Analysis and Machine Intelligence.* 24(7): 971–987.

[5] Liu, L., P. Fieguth, and G. Kauang. 2011. Generalized Local Binary Patterns for texture classification. In *Proceedings of the British Machine Vision Conference* (BMVA Press 2011) September: 123.1–123.11.

[6] Ma, Y. 2011. Number local binary pattern: An extended local binary pattern. *International Conference on Wavelet Analysis and Pattern Recognition* (ICWAPR 2011) July 978: 272–275.

[7] Guo, Z., L. Zhang, and D. Zhang. 2010. Rotation invariant texture classification using LBP variance (LBPV) with global matching. *Pattern Recognition* 43(3): 706–719.

[8] Guo, Z., L. Zhang, D. Zhang, S. Zhao, Y. Gao, and B. Zhang. 2010. A completed modeling of local binary pattern operator for texture classification. *IEEE Transactions on Image Processing* 19: 1657–1663.

[9] Zhao, S., Y. Gao, and B. Zhang. 2008. SOBEL-LBP. *IEEE International Conference on Image Processing* (ICIP 2008) October: 2144–2147.

[10] Zabih, R., and J. Woodfill. 1994. Non-parametric local transforms for computing visual correspondence. *European Conference on Computer Vision* (ECCV 1994) Springer-Verlag, New York 2: 151–158.

[11] Laws, K. I. 1980. Textured image segmentation. Image Processing Institute, University of Southern California, Report 940.

[12] Unser, M. 1995. Texture classification and segmentation using wavelet frames. *IEEE Transactions on Image Processing* 11: 1549–1560.

[13] Elnemr, H. A. 2013. Statistical Analysis of Law's Masks Texture Features for Cancer and Water Lung Detection. *IJCSI International Journal of Computer Science Issues* 10(6): 196–202.

[14] Setiawan, A. S., Elysia, J. Wesley, and Y. Purnama. 2015. Mammogram Classification using Law's Energy Measure and Neural Network. *Procedia Computer Science* 59: 92–97.

[15] Yadav, A. R., R. S. Anand, M. I. Dewal, and S. Gupta. 2015. Multiresolution local binary pattern variants based texture feature extraction technique for efficient classification of microscopic images of hard wood species. *Applied Soft Computing* 32: 101–112.

[16] Yadav, A. R., R. S. Anand, M. I. Dewal, and S. Gupta. 2015. Gaussian image pyramid based texture features for classification of microscopic images of hardwood species. *Optik* 126(24): 5570–5578.

[17] Tang, Z., Y. Su, M. J. Er, F. Qi, L. Zhang, and J. Zhou. 2015. A local binary pattern based texture descriptors for classification of tea leaves. *Neurocomputing* 168: 1011–1023.

[18] Dash, S., and U. R. Jena. 2017. Texture classification using steerable pyramid based Laws' Masks. *Journal of Electrical systems and Information Technology* 4: 185–197.

[19] Dash, S., and U. R. Jena. 2018. Multi-resolution Laws' masks based texture classification. *Journal of Applied Research and Technology* 15: 571–582.

[20] Dash, S., M. R. Senapati and U. R. Jena. 2018. K-NN based automated reasoning using bilateral filter based texture descriptor for computing texture classification. *Egyptian Informatics Journal* 19(2): 133–144.

Chapter 7

Analysis of BSE Sensex Using Statistical and Computational Tools

Soumya Chatterjee

Academy of Technology, Adi Saptagram, Hooghly 712121, West Bengal, India
Email: soumoch7@gmail.com

Indranil Mukherjee

Maulana Abul Kalam Azad University of Technology, West Bengal, Haringhata, Nadia 741249, West Bengal, India
Email: indranil.m11@gmail.com

7.1 Introduction 154
7.2 The Data Analysed 156
7.2.1 Return and Raw Data 156
7.2.1.1 Time series of the return data created from raw data 157
7.2.1.2 Return data created from detrended data . 158
7.2.1.3 The role of raw data in analyses 158
7.3 The Data Vectors and Principal Component Analysis 158
7.3.1 Construction of the Data Vectors 159
7.3.2 Principal Component Analysis 160
7.3.3 PCA of Raw Sensex Data 160
7.3.4 PCA of Detrended Sensex Data 161
7.3.5 PCA of Raw Sensex Data with Noise 161
7.3.6 PCA of Return Sensex Data 162
7.3.6.1 PCA of the return data obtained from raw Sensex data 162
7.3.6.2 PCA of the return data from detrended Sensex data 163
7.4 Kernel Principal Component Analysis 165
7.4.1 KPCA of Raw Sensex Data 165
7.4.1.1 Methodology 166
7.4.1.2 Results of KPCA applied to raw Sensex data (with trend) 166
7.4.2 KPCA of Raw Trend-Removed Sensex Values 166
7.4.3 KPCA of Raw Sensex Data with Noise 168
7.4.4 KPCA of Return Sensex Data 168
7.4.4.1 KPCA of return Sensex data 168

7.4.4.2 KPCA of return of detrended Sensex data 170
7.5 Detrended Fluctuation Analysis 172
7.5.1 Detrended Fluctuation Analysis of the Detrended Sensex Data 172
7.6 Conclusion 174
Bibliography 175

7.1 Introduction

Financial markets and financial data have been studied from different perspectives employing various types of tools. This is due to the growing importance of the financial indices as useful indicators of the economic health of a country [25]. However, equilibrium model and Efficient Market Hypothesis have failed to capture the essential characteristics of such markets including the explanation of the stylized facts [10]. One of the major reasons for this failure is the "complex" nature of financial markets, made up of different entities such as traders, speculators, hedgers, etc. interacting among themselves through non-linear mechanisms and trying to obtain the greatest profit [17].

A financial market is a system which seems to be perpetually in a non-equilibrium state and is driven by a large number of factors. It is therefore difficult to describe it analytically or to set up a mathematical model for it. However, the financial market, as a system, produces a huge volume of data in the form of financial indices. These data are recorded in real time (tick-by-tick). An analysis based on the available data can be made in which the signal (stock index) is studied to derive information about the system ("stock market") from which the signal emanates. It is in the analysis of this significant volume of data that techniques of statistical mechanics and computational sciences may be gainfully employed.

One aspect of the work focuses on the study of the correlations between the values of the stock index [26, 27] because the measure of the unsystematic correlations between the time series of stocks is important for capital allocation, risk management and portfolio management. Another facet of the analysis is to set up certain models that may facilitate the understanding of the portfolio and risk management procedures that are followed in finance [9]. The boundedness in the number of observations for each variable introduces estimation error or noise [6] in the analysis. This is the reason for imposing structures on the correlation matrix that will effectively reduce the dimensionality of the problem. Several statistical structures of the correlation matrices have been discussed in literatures [5] (and references therein).

Data from the Indian stock markets have been analysed to understand these markets. Owing to the high growth rate of its economy, India provides an ideal platform for such analysis. The analysis of BSE and NSE data revealed the existence of power law behaviour [22]. Another study reported that the changes in the Sensex were following an inverse cubic law [23]. The pattern observed in two different stock indices, viz. Nifty (NSE), and Hang Seng Index

(Hong Kong) was analysed in [14]. The existence of long-range behaviour and sensitive dependence on initial conditions were also presented [21].

In the present study, intraday prices are analysed during the period from 2006 to 2012 to gain an understanding about the nature of the Indian financial market. The objectives of the study can be delineated as follows:

1. To study the pattern of BSE Sensex and investigate the dimensionality of the system using using PCA and KPCA.

2. To obtain a qualitative understanding of the correlations using Hurst Exponents and Detrended Fluctuation Analysis.

It is to be noted that since principal component analysis (PCA) is a linear analysis technique while the financial market is, generally speaking, a highly non-linear environment, we are constrained to invoke the non-linear variant of PCA, the kernel principal component analysis (KPCA) to extract more accurate and meaningful results from the data being studied. The efficacy of the KPCA is evident from the fact that while PCA cannot identify the presence of noise in the signal (stock market), KPCA is able to do so.

Furthermore, it is necessary to detrend the Sensex data and capture purely the fluctuations present in the time series so that study of these fluctuations may help in understanding better the dynamics of the price data. Different methods are employed to remove trend in the data viz. First Differencing, Curve Fitting, Digital Filtering, etc. as well as adaptive mechanisms such as the empirical mode decomposition [16]. In this analysis, the time series data are detrended in a piecewise linear manner as per the following algorithm:

1. Identification of the breakpoints slowing slopes of the piecewise linear trends.

2. Fitting straight line segments between two successive breakpoints using least square approach.

3. Generate the detrended time series by subtracting from the given data the corresponding component obtained through straight line fitting.

In Figure 7.1, the original time series and the corresponding detrended data are plotted for the years 2006-2012. In Figure 7.2 the probability distribution functions are shown for the trend-removed data during 2006-2012. These distribution functions indicate departure from normality.

The uniqueness of the study is that the underlying dynamics governing the evolution of the stock market turns out to be correlated. This finding is in contradiction with the earlier assumptions that the governing processes are random and thus a random walk model could be used to explain its behaviour [3].

The remainder of the chapter follows the following structure. Section 7.2 discusses raw and return data and the nature of the autocorrelation functions obtained using these data. Section 7.3 elucidates how they are constructed and details of the various types of PCAs carried out using these vectors. Section 7.4

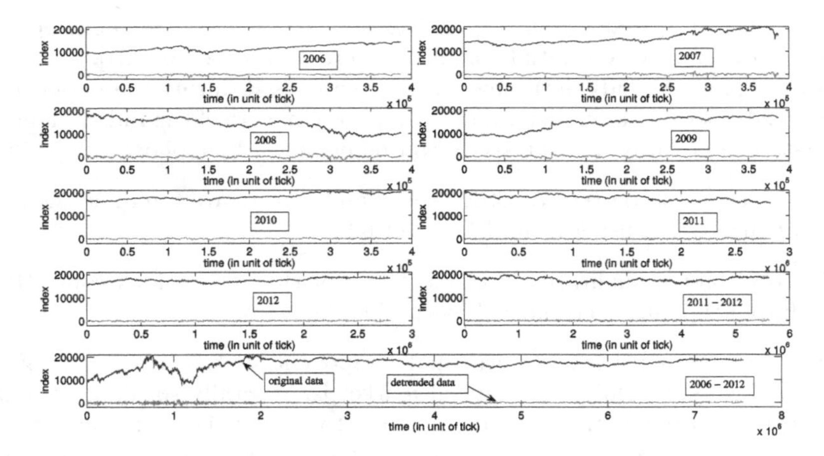

FIGURE 7.1: The plots show the original and the corresponding trend-removed data during the years 2006-2012.

presents the necessity of using KPCA and discusses the analysis undertaken using KPCA. Section 7.5 deals with the computation of the Hurst Exponent using Detrended Fluctuation Analysis and points out their relevance in the entire study. Section 7.6 summarizes our conclusions and alludes to the future scope of study.

7.2 The Data Analysed

The analysis is undertaken on the time series formed by using successive values of the BSE Sensex. Both raw and return datasets are considered. The returns are calculated from the original raw data as well as their detrended counterpart.

7.2.1 Return and Raw Data

To compare the returns over different time scales on an equal basis, logarithmic return has been frequently used in the statistical analysis of the prices [19, 20, 26]. Log transformation normalizes the variable so that all variables are mapped on to a comparable metric which allows evaluation of analytic relationships (e.g., covariance) between variables which originate from the price series of unequal values. Neither stock price nor its returns show normal distribution, but log returns being close to normal become more suitable for statistical analysis and forecasting. If $(P_1, P_2, ..., P_i, P_{i+1}, ...)$ are the price

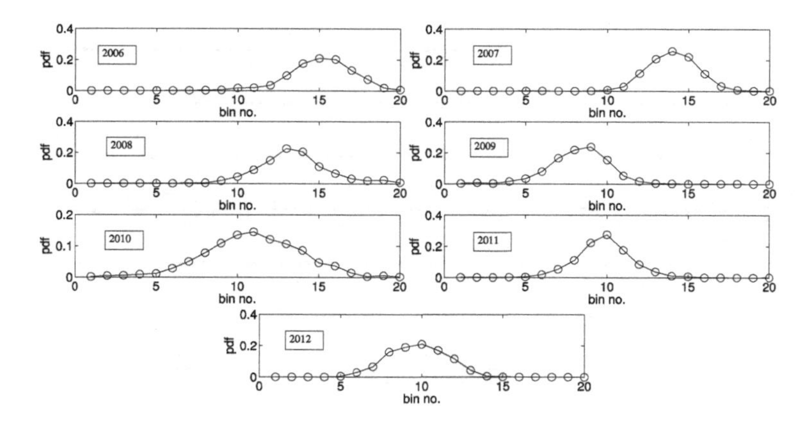

FIGURE 7.2: Plots of the probability distribution functions (pdfs) of the trend-removed data from 2006 to 2012.

values of a particular stock measured in increasing time, then the ratio P_{i+1}/P_i measures the fractional change in the price with time. This may be taken as a measure of instantaneous profit or loss of the particular price. The rate of return $\frac{P_{i+1}-P_i}{P_i}$ is used as another measure of profit and loss [4]. These different measures are used to find the short or long term correlations among the prices. It has been found through extensive studies of the prices over a sufficiently long time that over a short period of time there exists hardly any correlation in a particular price, but correlations start to emerge over longer time scales [11]. When the instantaneous fractional change in the price is very small, then in the first approximation $\log \frac{P_{i+1}}{P_i} \approx \frac{P_{i+1}-P_i}{P_i}$, which is the rate of return. The log return stabilizes the variance of a time series of price values. As a result, a substantial reduction is achieved in the mean squared error in forecasting using auto-regressive models. But as log transform is a highly non-linear transformation, the optimal forecast of an economic variable may not result in the optimal forecast of the transformed variable. The return of a log-return data may mask the internal mechanisms leading to the production of such data.

7.2.1.1 Time series of the return data created from raw data

The process of differencing which is a part of the methods that generate the return data makes the transformed data stationary even though not strictly stationary in general. The fluctuation in the magnitudes of the return data is thus expected to vary around zero value which seems to be the same as in Figure 7.3. However, the density of the spikes in the graphs varies with years and with time within a year as well.

The variation in the pattern of the intrinsic mechanisms driving the financial system may be considered to be the reason behind the variation in the peak density as a function of time.

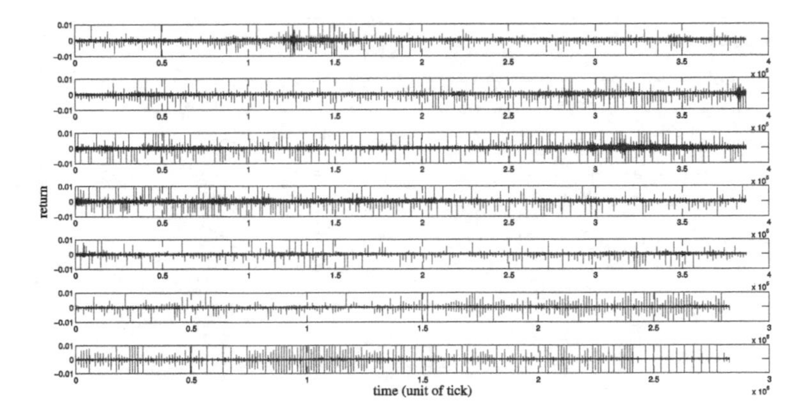

FIGURE 7.3: Graphs of the return data during 2006 to 2012. The fluctuations are registered around zero line. The pattern of the density of spikes changes with years and also with time within a year.

7.2.1.2 Return data created from detrended data

The nature of the graphs in Figure 7.4 shows features similar to those in the previous time series graphs. The magnitudes are found to fluctuate around the zero value and the density of the peaks keep changing with time. The only noticable change in the characteristics of the two graphs is the reduction in the peak number in any corresponding period. The reason for this change is the raw data which removes the spurious external perturbations from the system.

7.2.1.3 The role of raw data in analyses

It is felt that the objective of extracting the features and studying the pattern of evolution of the financial system becomes more meaningful only if the raw Sensex data or the detrended raw Sensex data is considered for study. The results arising from analysis of the raw data constitute a major part of this chapter, while the results coming from the analysis of return data manifest the difference in the statistical properties of the two datasets. The raw and the return datasets fall into two separate equivalence classes based on their symmetry properties.

7.3 The Data Vectors and Principal Component Analysis

This section discusses how data vectors can be constructed starting from the intraday BSE Sensex values. The analysis of these data vectors over a

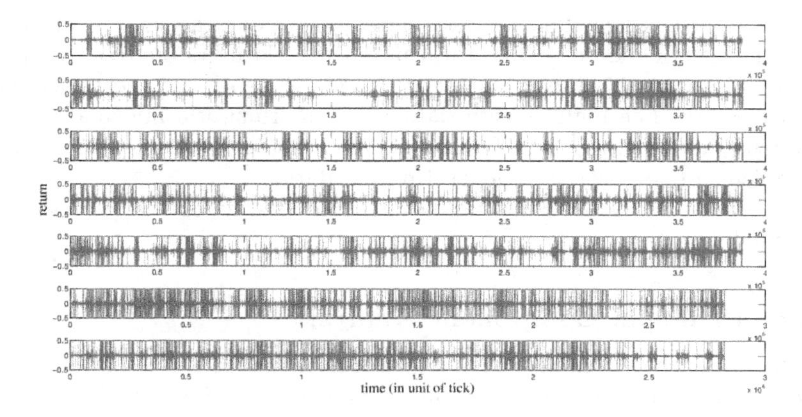

FIGURE 7.4: Graphs of the return of detrended data for the years 2006 to 2012. The fluctuations are registered around zero line. The pattern of the density of spikes changes with years and also with time within a year.

period of time gives an indication of the pattern in the Sensex data and thus an idea about the evolution of the financial market. PCA is applied to these data vectors. The datasets include the raw data and its detrended counterpart, raw data with noise, return data generated from the raw data as well as trend-free raw data.

7.3.1 Construction of the Data Vectors

Before considering the application of advanced statistical techniques viz., DFA, to the data it is necessary to suitably preprocess the data. The intraday prices for 2006 are taken and an interval of 50 minutes considered. This interval includes 200 data points wherein a 15-second gap is taken between successive prices. These prices are represented in the form of a column vector. $|x_1>= (P_1\ P_2\ P_3 ...\ P_{200})^T$. This window is definitely small relative to the time span of the entire year, thereby precluding the possibility of large price changes and allowing the use of a linear technique like PCA for the analysis. Such price vectors are constructed for the entire year 2006 by considering successive time intervals. Finally, 1800 vectors, each of dimension 200, are constructed. The collection of these objects represents the changes occurring in the index value during 2006. Since no analytical or qualitative description of price changes is possible, this methodology is a suitable alternative pathway to study the evolution of stock index.

Data for the five-year period 2006-2010 are considered individually on a yearly basis and also cumulatively over periods of five to two years, thereby creating 15 datasets. The cumulative break-up of the data of 2011 and 2012 has been done similarly but analysed separately due to the significant increase in the density of the data. The total number of datasets analysed is thus 18 (15+3). The analysis of these datasets of different densities nevertheless yields similar results. The high density data yields more subtle information about the intrinsic structures of the Sensex prices which leads to finer details of the mechanisms of the processes generating the data.

The data vectors are constructed in a similar way for the analysis of the detrended raw data.

7.3.2 Principal Component Analysis

The presence of a number of agents in the stock market could lead to such systems possessing high dimensionality. However, the effective dimensionality can be gauged using PCA. PCA transforms the original variables to a set of uncorrelated but ordered variables, called the principal components, wherein the first few components reflect most of the variations contained in the original variables [24, 28]. This is done through rotation of the original basis.

However, one critical limitation of PCA is the linear nature of the transformation, in sharp contrast to actual processes which are not so [8].

In PCA, under rotation, the initial vector space goes over to a new one having the basis vectors $|\phi_i>$ in such a way that in the new coordinates, the variance retained by the data is maximized [2, 13, 18]. The following steps are followed:

(a) The covariance matrix is determined:

$$\Sigma_x = \frac{1}{N}\sum_{i=1}^{N}(|X_i> - |\bar{X}>)(<X_i| - <\bar{X}|) \tag{7.1}$$

where N denotes the total number of vectors and $|\bar{X}>$ is their mean.

(b) The eigenvalues of the covariance matrix and the eigenfunctions belonging to them are obtained. The eigenvectors $|\phi_i>$, which are mutually orthogonal and individually normalized, form a complete set. The kth Principal Component (PC) is given by $y_k =<\phi_k|X>$ where $|\phi_k>$ is an eigenvector of Σ_x corresponding to the kth largest eigenvalue λ_k.

7.3.3 PCA of Raw Sensex Data

The 18 data systems are subjected to PCA. The eigenvalue spectrum indicates that in every case, an amount of trace exceeding 99% is consumed by the first eigenvalue. Figure 7.5 shows the graphs of the normalized eigenvalues for the individual years from 2006 to 2012 as well as for the cumulative dataset covering the years 2006-2012.

Each eigenvector corresponding to the highest eigenvalue has dimension 200. All possible angles among them, numbering C_2^{18} or 153, are nearly zero, indicating that these eigenfunctions have nearly the same direction.

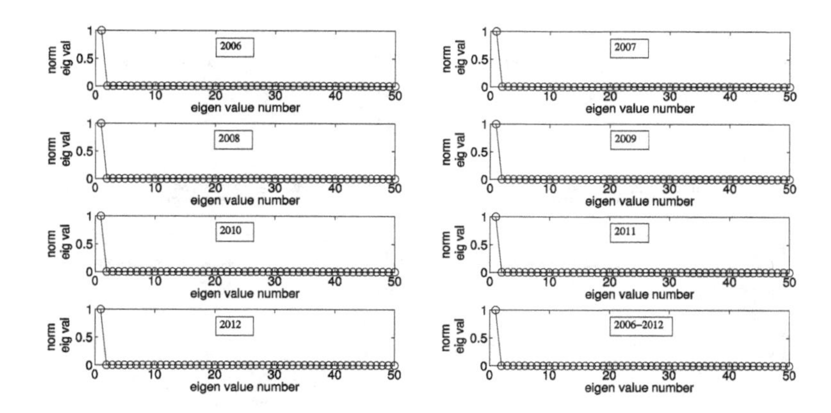

FIGURE 7.5: Graphs of normalized eigenvalues against serial number for the individual years from 2006 to 2012 as well as for the cumulative data covering the entire time period, using PCA.

These two facts, viz. dominance of one single eigenvalue and unidirectionality of the eigenvectors, irrespective of the year number or the data size, allude to high degree of correlation in the data. A very small subspace of the initially constructed vector space of 200 dimensions suffices to represent the majority of the data in an optimal fashion. The results continue to hold good when the dimension of the vector space is changed.

7.3.4 PCA of Detrended Sensex Data

PCA is applied on detrended Sensex data by constructing data vectors as described above and using vector spaces of different dimensions. The first one or two eigenvalues are always found to carry 99% of trace.

Figure 7.6 shows the eigenvalue vs. serial number graph using trend-removed Sensex prices for 2007, containing four different subgraphs showing usage of vector spaces of differing dimensions. Similar plots will be obtained when data for other years are considered. On the whole, it is clear that the behaviour of the detrended data is quite similar to that of the one carrying trend.

7.3.5 PCA of Raw Sensex Data with Noise

To examine whether external perturbation has any effect on the index values, noise is introduced at different levels in the data. To mix noise of 10% magnitude to a stock price of 22,000 (*say*), 10% of 22,000, i.e., 2200

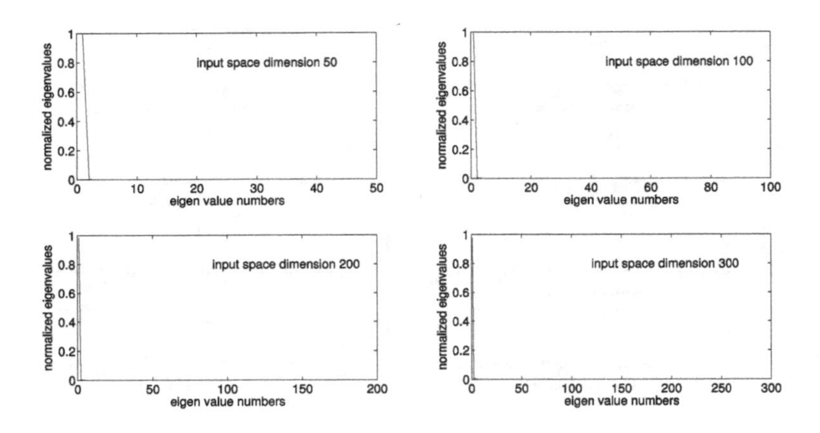

FIGURE 7.6: Graph of the eigenvalues versus serial number of the trend-removed prices for 2007 under PCA, using spaces of differing dimensions.

is multiplied by any random number lying between 0 and 1 and this value is added to 19,800. The stock price is now composed of 90% of the original value and 10% of perturbation. The process is repeated for all possible prices, thereby adding 10% noise to the data. But application to the noisy data does not indicate any departure from the original result, possibly owing to the linear characteristics of PCA.

7.3.6 PCA of Return Sensex Data

PCA is applied to the 18 return datasets obtained from both the original raw datasets and the detrended datasets. Feature extraction or the effective dimensionality identification is not the primary objective of this analysis because the return data possess non-linearity of high degree. The non-linearities result from the complex mechanisms driving the financial system as well as from the non-linear transformation that produces the data. The analysis is carried out following the same prescription as discussed in the previous sections, i.e., by creating data vectors of dimensions 50, 100, 200, 300. The selection of these dimensions are based on the principle of coarse graining so that the vital information regarding the pattern of the dynamics of the system is preserved.

7.3.6.1 PCA of the return data obtained from raw Sensex data

The results of the analysis for the individual years show that there is hardly any signature in the return data. Moreover, the pattern of the results remains almost the same irrespective of the dimension of the input vectors. This observation is quite evident in Figure 7.7. However, the comparison of

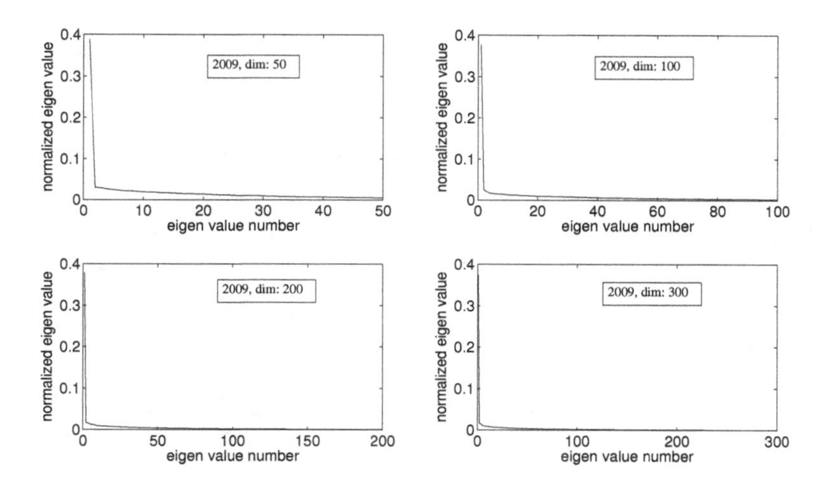

FIGURE 7.7: Graph of the eigenvalues versus the serial number for return data of 2009 obtained by applying PCA to vector space of varying dimensions.

the pattern of the result for different years manifests that even though the charateristics of low correlation is common in the return datasets, there is discernible variation in the pattern for different years. This observation is distinct in the graphs of Figure 7.8.

The variation in the PCA results with respect to the year number may be attributed to the fact that the degree of influence of external factors on the intrinsic dynamics of the financial system changes with time.

7.3.6.2 PCA of the return data from detrended Sensex data

The results of the analysis of return data produced from the detrended raw Sensex data do not seems to change much compared to that of the return data created from the original (non-detrended) raw data. The pattern in the results is not influenced by change in the input dimension for a particular time interval (in the units of months or units of years) which can be seen in Figure 7.9. But there exists a variation in the pattern of the results for different time intervals, i.e., for different years. This variation can be observed in Figure 7.10. But the comparative study of the PCA results for the two types of return data shows that the number of eigenvalues which accounts for the 90% of the trace, for example, is less in case of the returns obtained from the detrended Sensex values. This is true for all the years considered for study.

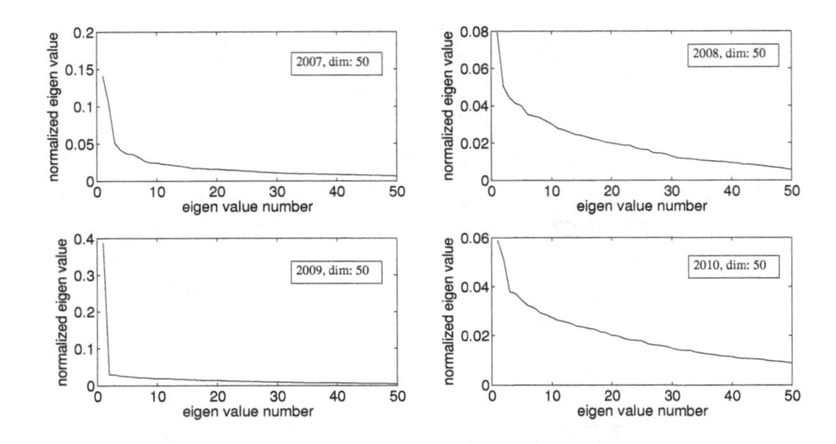

FIGURE 7.8: Graph of the eigenvalues versus the serial number for the return data of the years 2007-2010 calculated for input dimension 50 using PCA.

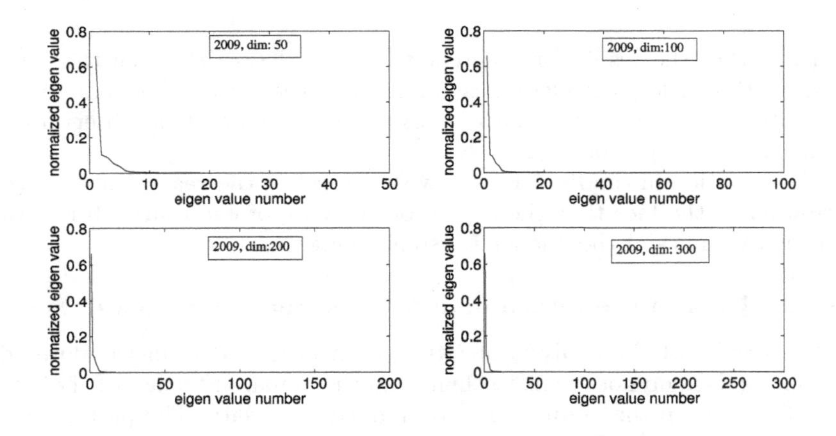

FIGURE 7.9: Graph of eigenvalues versus serial number for returns obtained from trend-removed Sensex prices of 2009 for dimensions 50, 100, 200 and 300 using PCA.

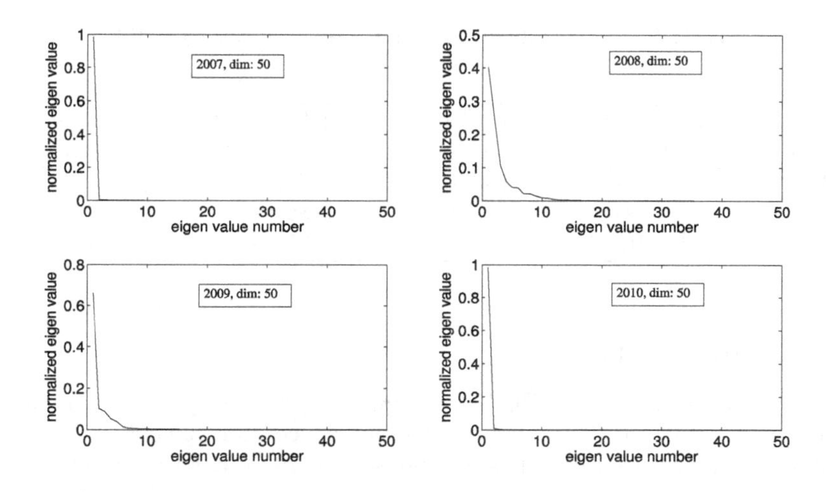

FIGURE 7.10: The plot reveals that the first few eigenvalues do not consume any significant trace pointing to low correlation. This happens for all the years but amount of correlation varies from year to year.

7.4 Kernel Principal Component Analysis

PCA is unable to detect non-linear variations in the dataset since it is essentially a linear technique. It calls for a higher order non-linear variant of PCA which will extract the non-linear features in the data. This role is fulfilled by KPCA. Even when there are non-linear variations in the data, it can be mapped to a higher dimensional space, called Feature Space, wherein the variation becomes linear in nature. The mapped data is subjected to PCA. The Feature Space F may have a very high dimension. This happens when the mapping is done into the space of all possible k^{th} order monomials in input space. Without directly carrying out the mapping Φ, KPCA uses Mercer kernels to achieve the same objective. Finally, KPCA boils down to the solution of an ϵ value problem and usage of different types of Kernels ensures that non-linearities of different classes can be addressed using KPCA.

7.4.1 KPCA of Raw Sensex Data

KPCA is used on the time series constructed by using raw BSE Sensex data. The following subsections give the details.

TABLE 7.1: Application of KPCA on dataset of 2007.

Dimension of input space	No. of vectors	First eigenvalue	First five eigenvalues	First ten eigenvalues
200	1800	59.03	87.48	94.85
100	3600	55.59	89.87	97.45
50	7200	61.58	95.18	98.76

7.4.1.1 Methodology

KPCA implements the PCA technique in Feature Space instead of the input space. With a 200-dimensional input space and 390,000 prices being analyzed, the Feature Space has a dimension of 1950. For the mapping function, polynomial kernel of different powers is used, starting from 2, the first non-trivial higher order polynomial.

7.4.1.2 Results of KPCA applied to raw Sensex data (with trend)

The result of the applied KPCA to the raw Sensex data is shown in tabular form.

Table 7.1 shows the results upon applying KPCA to data of 2007 using kernel of power 2. Variation in the dimension of input space and number of vectors and the resultant % of trace captured by the first one, the first five and the first ten eigenvalues are displayed. Similar exercise carried out with polynomials of different higher powers shows that 95% of the trace is consumed by the first ten eigenvalues. Reduction in input space dimension and the number of vectors leads to higher percentage of trace accounted for by the leading eigenvalues. This condition leads to increase in the dataset because of the reduced time difference in capturing of the prices. Since the feature space has a greater dimension compared to original input space, KPCA also indicates high correlation in the prices.

Figures 7.11 shows the results of applying KPCA to index values of 2008.

7.4.2 KPCA of Raw Trend-Removed Sensex Values

When KPCA is applied to detrended Sensex data, once again a polynomial Kernel is employed with the power increasing from 2 onwards.

The analysis is done on trend-removed data for years 2006 to 2012. Figure 7.12 displays the graph of eigenvalues versus serial number for the year 2006. The four subgraphs in Figure 7.12 show the analysis done with vector space of varying dimensions.

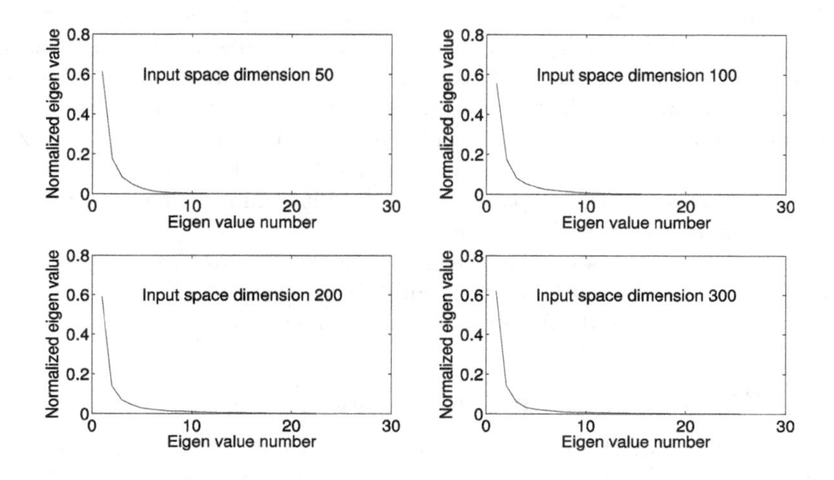

FIGURE 7.11: Graph of eigenvalues versus serial number for index values of 2008 upon applying KPCA.

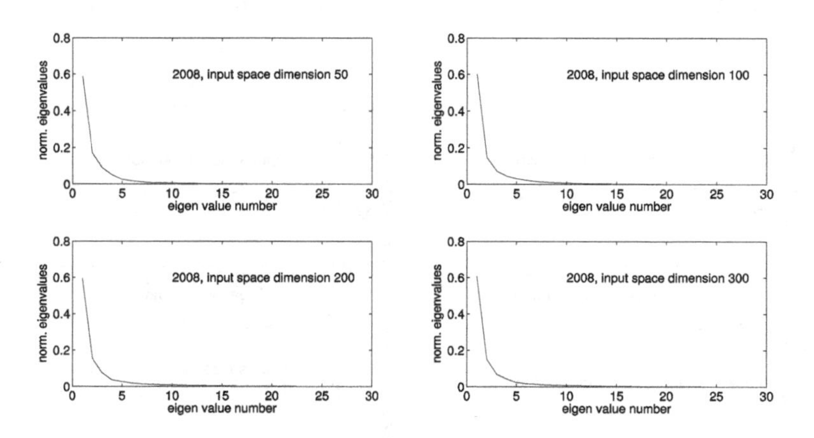

FIGURE 7.12: Graph of eigenvalues versus serial number of the detrended Sensex prices for 2006 applying KPCA. The spectrum shows the first few eigenvalues accounting for almost the entire trace.

7.4.3 KPCA of Raw Sensex Data with Noise

To analyse the effect of external perturbation on index values when non-linearity is considered, KPCA is used on the index values of 2008 with varying levels and a polynomial kernel of order 2. Variation in the input space dimension and computation of the results show a distinct fall in the percentage of the trace being captured relative to the noise-free situation, indicating lowering of the data upon contamination by external noise. It may be mentioned here that PCA could not detect this departure. This result points to the significance of non-linearity in understanding of the system.

Figures 7.13 and 7.14 display the eigenvalue spectra derived from applying KPCA on index values of 2008 with addition of noise at 5% and 10% levels, respectively. Each case displays the leading 30 eigenvalues.

7.4.4 KPCA of Return Sensex Data

In the KPCA of the raw data it is known that the first 10 eigenvalues capture almost 98% of the trace irrespective of the data size, but KPCA of the return data shows quite a variation in the number of eigenvalues capturing such amount of trace.

7.4.4.1 KPCA of return Sensex data

If the number of eigenvalues n capturing 98% of the trace is taken as a parameter to measure the degree of correlation, then Figure 7.15 shows that in case of 2010 the number n turns out to be quite less compared to that

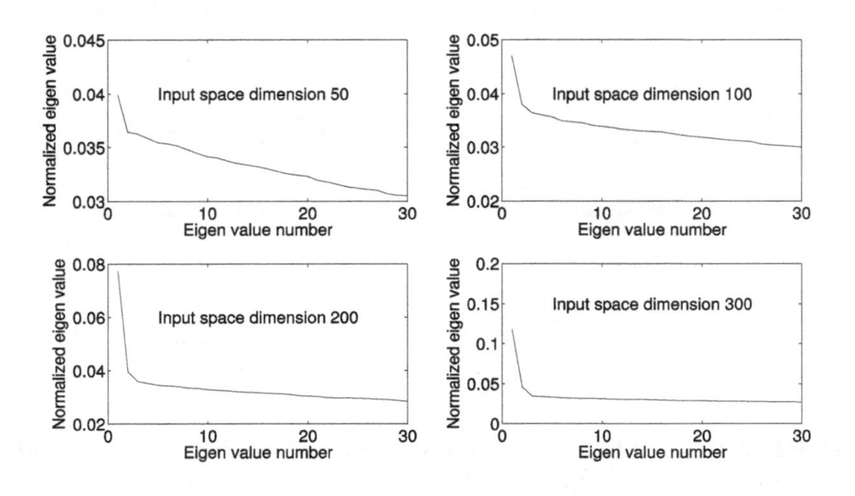

FIGURE 7.13: Graph of eigenvalues versus serial number for index values of 2008 with noise at 5% level, derived by applying KPCA of order 2.

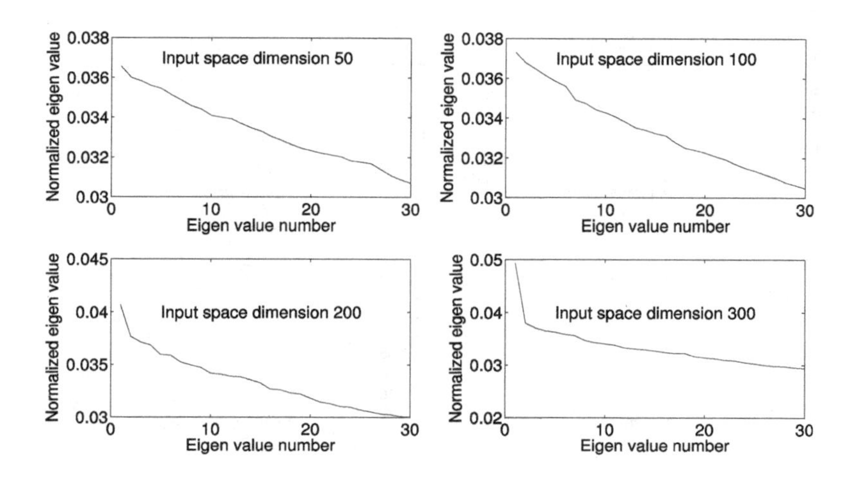

FIGURE 7.14: Graph of eigenvalues versus serial number for index values of 2008 with noise at 10% level, derived by applying KPCA of order 2.

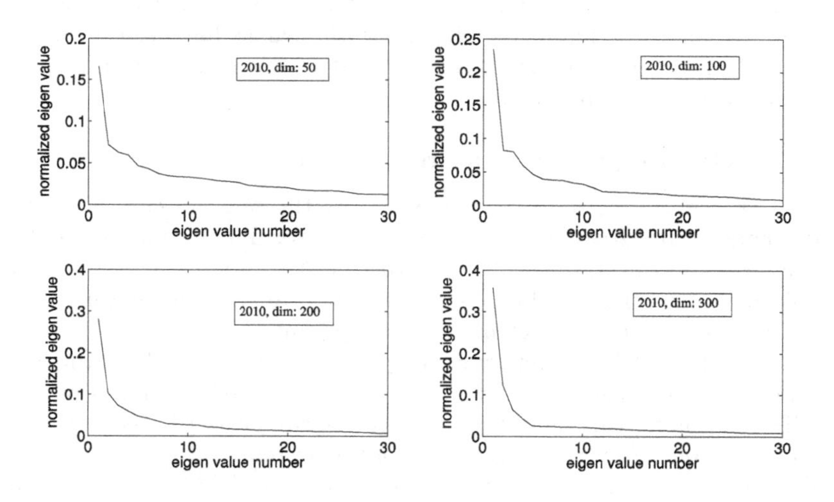

FIGURE 7.15: Graph of eigenvalues versus serial number for return data of 2010 applying KPCA.

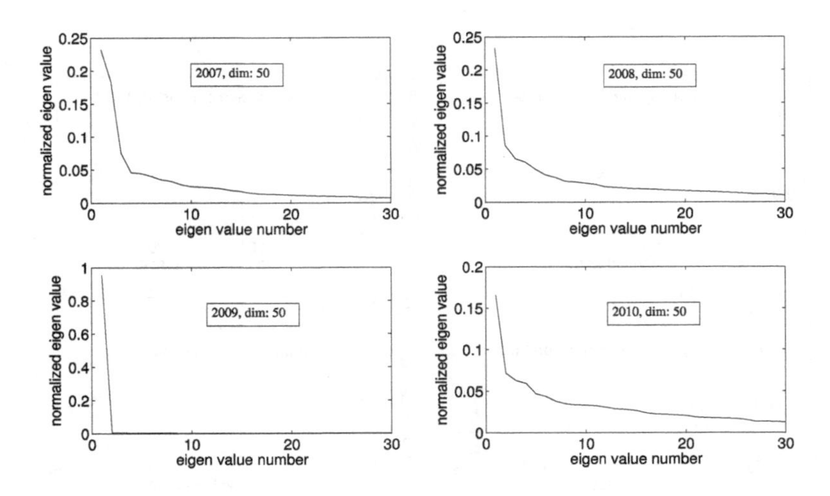

FIGURE 7.16: Graph of eigenvalues versus serial number for return of the raw Sensex data from 2007 to 2010 applying KPCA. The spectrum indicates highly correlated behaviour of the return data. There are distinct variations in the pattern of KPCA for different years.

obtained from PCA. This particular characteristic is universal across all the years even though the number n changes with year as has been displayed in Figure 7.16. But for each year this number does not change with respect to the dimension of the input vector.

7.4.4.2 KPCA of return of detrended Sensex data

The features of the results remains almost the same in case of return of the detrended data. Irrespective of the input vector dimension, the number of eigenvalues capturing the 98% trace n turns out to be low in comparison to that obtained from PCA. This observation is quite clear in Figure 7.17. But there are hardly any visible changes in the parameter n, i.e., no discernable variation in the KPCA pattern with respect to year as manifested in Figure 7.18. This is in contrast to the result observed in case of return of raw Sensex data. The reason for this significant observation can be attributed to the effect the trend has on the intrinsic dynamics of the system.

The effective dimensionality of the feature space in KPCA being arbitrarily large, the numerical change in the ratio of n to the large effective dimension turns out to be negligibly small. This degree of smallness is not taken as a measure of the variation in the number n as observed in the results of the KPCA in case of raw and return data in our analysis. But the reduction in n explores the qualitative difference in the statistical properties of the raw and the return data.

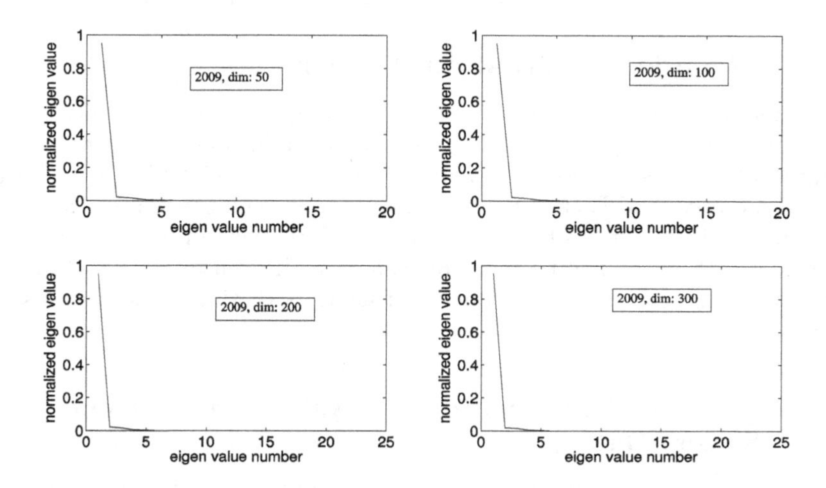

FIGURE 7.17: Graph of eigenvalues versus serial number for return of the trend-removed Sensex data of 2009 applying KPCA. The spectrum indicates highly correlated behaviour.

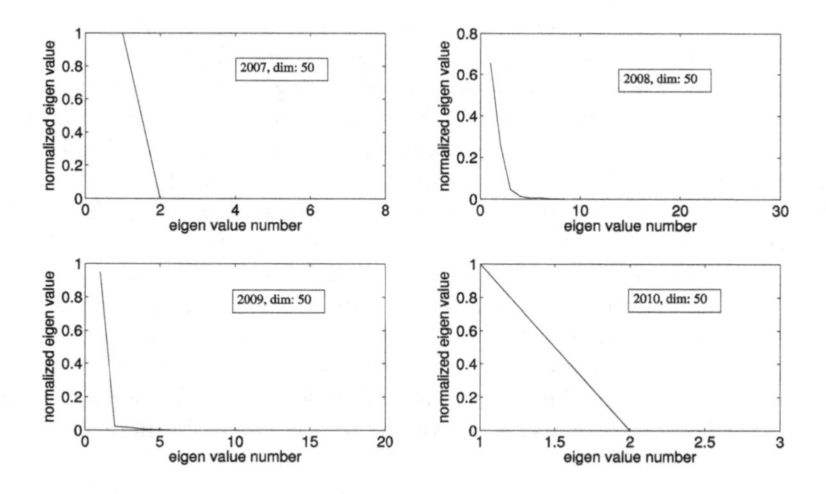

FIGURE 7.18: Graph of eigenvalues versus serial number for return of the trend-removed Sensex values from 2007 to 2010 applying KPCA. The spectrum indicates highly correlated behaviour of the return data.

7.5 Detrended Fluctuation Analysis

Detrended fluctuation analysis (DFA) technique has been applied to find the degree of scaling in the data but with an additional advantage of its applicability in case of non-stationary data [1, 7, 12, 15, 30]. Tests like the Philip-Peron test and Dickey-Fuller test have confirmed the non-stationarities of the data which justifies the continuous changes occurring in the market on the whole with time. The presence of scaling in the Sensex data for different years and cumulatively have been found using both the Hurst analysis and DFA.

7.5.1 Detrended Fluctuation Analysis of the Detrended Sensex Data

DFA is supposed to be a better candidate for finding the scaling/persistence/memory in the data taking the non-stationary property of the Sensex data into consideration. The analysis of Sensex data reveals the value of the scaling exponents $\alpha = DFA_{Hurst}$ to be around just below 1.5 for each individual year as displayed in Figure 7.19. Figure 7.20 indicates that the value of α calculated on the cumulative datasets of two consecutive years also lies within the same range. There is a universality in the value of α with respect to the different intervals of time chosen. The collinearity of points on the log-log graph is indicated across a sufficiently large range of window sizes. Like the results of the individual years, the range of α for cumulative years also lies between 1.40 to 1.46. So, the possibility of finding multi-scales for the datasets analysed covering from 2006 to 2010 can be nullified. It had been found by Kantelhardt et al. [29] that the scaling exponent calculated using DFA can be interpreted as a kind of generalization of the classical Hurst exponent. It provides a quantitative measure of the scaling behavior in the fluctuations of the second moment after detrending. The classical Hurst exponent corresponds to the second moment for stationary cases and to the second moment minus 1 for non-stationary cases. The scaling exponent value of 1.5 is thus equivalent to 0.5 for non-stationary cases. The scaling exponents calculated by two methods are thus close to each other. How effectively DFA removes the non-stationarities of the data is still questionable. This limitation in the DFA can be considered as a plausible reason for whatever small difference remains in the scaling exponent values calculated by two methods.

It is quite clear from the figures that the calculated values of the fluctuation lies very close to the fitted straight line for all the graphs. This feature is the primary characteristics of the self-similarity. This characteristics of self-similarity found in the datasets points to the fact that the value α is a fractal dimension which shares all the desirable properties of the Hausdorff dimension. The fractal nature of the datasets as found using DFA supports the similar

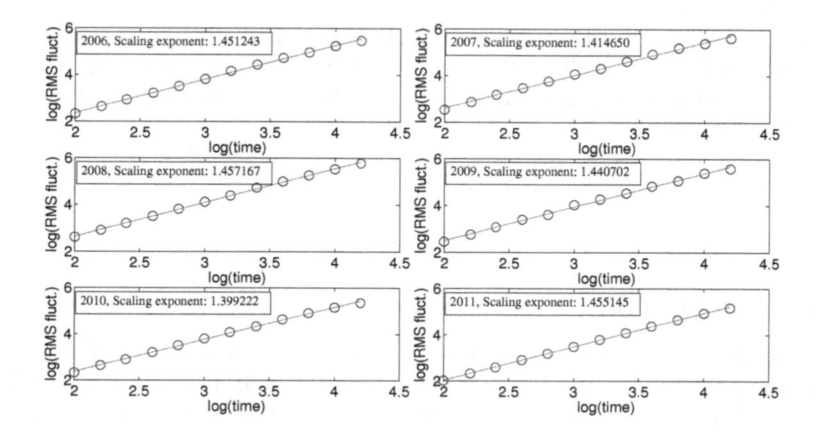

FIGURE 7.19: Plot of the Hurst exponent (DFA_{Hurst}) against number which corresponds to increasing time intervals in log-log scale for the years 2006 to 2011 using DFA technique. The collinearity of points on the log-log graph is indicated across a sufficiently large range of window sizes.

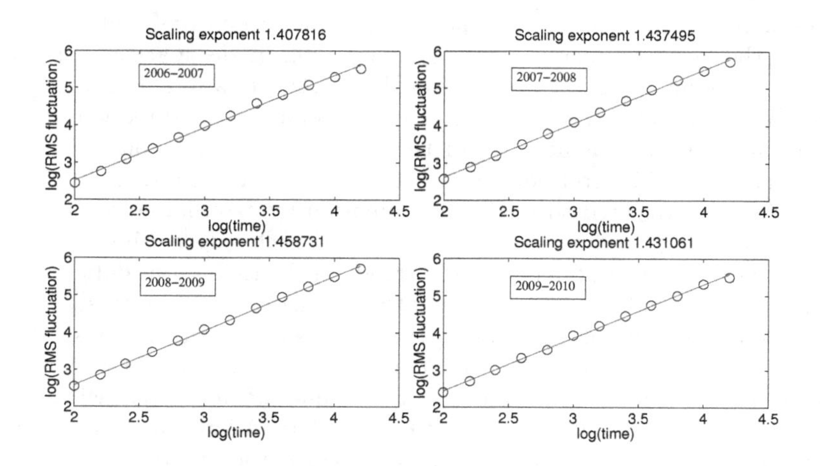

FIGURE 7.20: Plot of the Hurst exponent (α) against number which corresponds to increasing time intervals in log-log scale for the years 2006 to 2011 by taking data of two years cumulatively using DFA technique.

findings of the datasets to be fractal based on the sliding window technique of Hurst analysis.

The exponent α of Fractional Brownian Motion (FBM) lies between 1 to 2. This similarity in the value of α between FBM and Sensex data points towards the possibility of modelling the dynamics of Sensex data using FBM stochastic models.

7.6 Conclusion

The aim of this work was to understand the dynamics of the Indian stock market using Sensex values as indicator of the market behaviour. It is extremely difficult to achieve this objective given the huge number of agents in the market and their non-trivial interactions. Tools drawn from advanced statistical mechanics and computational sciences were utilized to achieve our objectives. The importance of KPCA arises from the non-linearity of the interactions driving the market. The impact of external perturbation on the market index has been modeled by mixing of different magnitudes with the index values ($the signal from the market$). In almost all situations considered, the first few eigenvalues account for an overwhelming bulk of the trace. This reflects a high degree of correlation in the Sensex values and indicates that the dynamics is restricted to a small subspace of the initial vector space where the analysis is initiated. It is also apparent that the presence of and interactions among the large number of agents participating in the market activities do not alter the macro-level properties of the market, in any significant manner. Despite fluctuations, the market gets steered in a specific direction. It is only significant external disturbance in the form of any socio-political upheaval that may alter this situation in a non-trivial manner. PCA fails to discern the difference triggered through contamination by external noise, but KPCA shows the change in data behaviour caused by such external interference.

It is still a matter of debate if the trends in the time series data are due to some external conditions or are generated as part of the intrinsic dynamics of the system. Detrending of the data therefore becomes the obvious extension of the previous work to avoid this ambiguity. The detrended data represents a time series generated from the intrinsic dynamics of the system which has no influence of the external conditions. PCA and KPCA have been applied to the detrended data to observe any change in the correlation properties of the data. The need to acquire a more detailed information about the correlation found in the data in previous analysis crops up from the need to find if the external perturbations present in any financial systems have any direct influence on the correlation properties of the data.

The study of the trend-removed intraday Sensex data covering the period 2006-2012 was intended to unearth patterns in the data, if any. The detrending led to removal of non-stationarity in the data and left fluctuations about a zero-mean process. The analysis revealed the following major features of the detrended data:

(i) The existence of a high degree of correlation was revealed through application of PCA and KPCA.

(ii) A fairly high degree of persistence was revealed upon computation of Hurst exponents using DFA, both on an individual as well as cumulative basis.

(iii) The analysis was done for quite large datasets over a period spanning from 2006 to 2012. The repetitive and reproducible behaviour observed through analysis reflects robustness of the process/system (the financial market) producing the data.

(iv) The time span under analysis covers a certain period of global recession and it was essential to look for any signature of impact of the recession on the Indian economy, since stock indices can be regarded as some barometer of the economy. The findings did not indicate any visible impact of the downturn on the parameters analysed as part of this study.

Bibliography

[1] J. Alvarez-Ramirez, J. Alvarez, E. Rodriguez, and G. Fernandez-Anaya. *Physica A*, 387:6159, 2008.

[2] S. Axler. *Linear Algebra Done Right.* Springer, 1997.

[3] F. Black and M. Scholes. *Jour. of Political Economy*, 81:637, 1973.

[4] Z. Bodie, A. Kane, and A.J. Marcus. *Essentials of Investments.* McGraw-Hill/Irwin, 2004.

[5] L. Chan, J. Karceski, and J. Lakonishok. *Rev. Fin. Stud.*, 12:937, 1999.

[6] V. Chopra and W.T. Ziemba. *J. Portf. Manage.*, 19:6, 1993.

[7] R.L. Costa and G.L. Vasconcelos. *Physica A*, 329:231, 2003.

[8] D. Dong and T.J. McAvoy. *Computers & Chemical Engineering*, 20(1):65, 1996.

[9] E.J. Elton and M.J. Gruber. *Modern Portfolio Theory and Investment Analysis.* Wiley, New York, 1995.

[10] E.J. Fama. *J. Finance*, 25:383, 1970.

[11] Justin Fox. *Myth of the Rational Market.* Harper Business, 2009.

[12] D. Grecha and Z. Mazur. *Physica A*, 336:133, 2004.

[13] G.Strang. *Linear Algebra and Its Application.* Academic Press, 1976.

[14] K. Guhathakurta, I. Mukherjee, and A. Roy Chowdhury. *Chaos Solitons & Fractals*, 37:1214, 2008.

[15] R. Hardstone, S.-S. Poil, G. Schiavone, R. Jansen, V.V. Nikulin, D. Mansvelder, Huibert, and K. Linkenkaer-Hansen. *Physica A*, 3, 2012.

[16] N.E. Huang, Z. Shen, S.R. Long, M.C. Wu, H.H. Shih, Q. Zheng, N.C. Yen, C.C. Tung, and H.H. Liu. *Proc. R. Soc. Lond. A*, 454:903, 1998.

[17] N.F. Johnson, P. Jefferies, and P.M. Hui. *Financial market complexity: What Physics Can Tell Us about Market Behaviour.* Oxford University Press, 2003.

[18] I.T. Jolliffe. *Principal Component Analysis.* Springer, 2002.

[19] L. Laloux, P. Cizeau, M. Potters, and J.-P. Bouchaud. *Int. Jour. of Theor. and Appl. Fin.*, 3:391, 2000.

[20] H. Lütkepohl and F. Xu. *Empirical Economics*, 42:619, 2012.

[21] R.K. Mishra, S. Sehgal, and N.R. Bhanumurthy. *Rev. Finance Econ.*, 20(2):96, 2011.

[22] R.K. Pan and S. Sinha. *Econophysics of stock and other markets:A.Chatterjee, B.K.Chakraborti(Eds.).* Springer, Milan, 2006.

[23] R.K. Pan and S. Sinha. *Physica A*, 387:2055, 2008.

[24] D.J. Patil, B.R. Hunt, E. Kalnay, J.A. Yorke, and E. Ott. *Phys. Rev. Lett.*, 86:5878, 2001.

[25] K. Pilbeam. *Finance and Financial Markets.* Macmillan Business, 1998.

[26] V. Plerou, P. Gopikrishnan, B. Rosenow, L.A.N. Amaral, T. Guhr, and H.E. Stanley. *Phys. Rev. E*, 65:066126, 2002.

[27] V. Plerou, P. Gopikrishnan, B. Rosenow, L.A.N. Amaral, and H.E. Stanley. *Phys. Rev. Lett.*, 83:1471, 1999.

[28] J.T. Scheick. *Linear Algebra with Applications.* McGraw-Hill, New York, 1997.

[29] H.E. Stanley, V. Kantelhardt, S.A. Zschiegner, E. Koscielny-Bunde, S. Havlin, and A. Bunde. *Physica A*, 316:87, 2002.

[30] Y. Wang, L. Liu, R. Gu, J. Cao, and H. Wang. *Physica A*, (389):1635, 2010.

Chapter 8

Automatic Sheep Age Estimation Based on Active Contours without Edges

Aya Abdelhady

Cairo University, Egypt

Aboul Ella Hassanien

Faculty of Computers and Information, Cairo University, Egypt
E-mail: aboitcairo@gmail.com

Aly Fahmy

Cairo University, Egypt

8.1 Introduction 177
8.2 Related Work 178
8.3 Theory and Background 180
 8.3.1 Active Contours 180
 8.3.2 Blob Detection and Counting 181
 8.3.3 Morphological Operations 181
 8.3.4 Dentition 181
 8.3.5 Image Collection and Camera Setting 181
8.4 The Proposed Automatic Sheep Age Estimation System 182
 8.4.1 Pre-processing Phase 183
 8.4.2 Segmentation Phase 183
 8.4.3 Post-processing Phase 185
 8.4.4 Age Estimation Phase 185
8.5 Experimental Results and Discussion 188
8.6 Conclusion and Future Work 192
 Bibliography 192

8.1 Introduction

Sheep age is a vital feature in animal management. Decisions of when to cull and when to mate are highly dependent on age. Sheep age is also important in the food market, as it affects the quality of food because older sheep usually have darker and less tender meat [22]. In addition, most of the sheep are kept in farms and it is usually hard in farms to keep records of animal ages. Moreover, it is difficult to identify the sheep. Farmers in Egypt identify the sheep using ear tags. However, numbers on ear tags are usually washed away and ear tags often fall from the sheep's ear. As a result, farmers know sheep age by counting their teeth and estimating their size. Number and size of teeth are used worldwide to calculate the age of cattle. This approach is known as dentition. However, most of the consumers do not have that experience to detect the age of animals by dentition. Therefore, they can get manipulated by sellers. Moreover, farmers are prone to mistakes. In addition, most of the farmers do not have the same opinion about a sheep's age range. Sheep age does not have to be calculated precisely in months and days. However, it is enough to detect age ranges to take important management decisions. Therefore, in this chapter, a real time application which estimates age ranges of sheep by analyzing only one image for teeth is proposed. As mentioned before, age can be estimated for all ruminant animals in the same way as sheep [26]. For this reason, the proposed approach in this chapter can be used to estimate the age of other ruminant animals. Teeth were segmented from images using active contours. Active contours are well-known methods used for image segmentation. Active contours work by setting initial contours, and then these contours evolute in each iteration. The contours keep shrinking or expanding by minimizing a defined energy function. Then, number of teeth and their linear dimensions were determined from segmented teeth images using visual analysis techniques. Age was estimated from teeth count and length because younger sheep have milk teeth which are very short and very small. On the other hand, older animals have permanent teeth which are taller and they also count more. Therefore, age is determined by the number of pairs of permanent teeth. Furthermore, a broken tooth is an indication of aged sheep [3, 26]. According to literature survey, age estimation of sheep has received inadequate attention. Moreover, researches usually used radiography for animal age estimation [7, 15, 16, 20]. However, the proposed approach is the first to estimate a sheep's age by analyzing images taken by a mobile camera which is available for all producers and consumers. All of the work conducted in this chapter was on data collected for 52 Barki sheep.

The remainder of this chapter is organized as the following. In Section 8.2, related work is briefly mentioned. Section 8.3 explains important theoretical background. Then in Section 8.4, the proposed approach for age estimation is discussed in details. Further, in Section 8.5, experimental results are presented and evaluated. At last, conclusion and future work directions are provided in Section 8.6.

8.2 Related Work

There have been already several approaches for human age estimation which is more complicated than approaches for animals' age estimation because humans have several age ranges. Moreover, humans have many teeth in both the lower and the upper jaws. There were multiple methods tested on humans for age estimation which were basically: Morphological methods, biochemical methods and radiological methods [5]. Morphological methods required tooth extraction for microscopic assessment which was against ethics. Moreover, it needed huge effort. The biochemical methods used racemization of amino acids, as they were faster in cells of lower metabolism. This approach was also time consuming and needed special equipment. Radiology was also used for age detection, but it also needed special tools to obtain these radiological images. On the other hand, for age estimation of animals, many techniques have been examined. Some techniques were cruel to animals because they also needed tooth extraction to determine the age. Tooth had to be extracted to count cementum lines for age estimation of lions because their count represents age ranges [30]. Counting cementum lines were also used for age estimation of other animals such as coyotes and gray wolves [8, 19, 28]. However, tooth extraction for age estimation of living animals is very cruel and ethically unacceptable [4, 17]. Moreover, counting cementum lines was time consuming and required special tools. Other approaches required sensitive nucleus or ocular lens examination which also needed special atmosphere [11, 23]. For age estimation of cats and dogs, they had to be locked in dark room and tested with penlights. Younger cats and dogs have bigger lens reflections in their eyes. Moreover, lens turns to blue as animals age. These features were used to categorize cats and dogs into different age ranges. Dental radiographic images were also used to monitor the changes in pulp cavity size and analyze the relation with the age of different animals. This methodology was accurate as teeth are not affected by hormonal changes. Moreover, teeth were not extracted because sensors were inserted inside animals' mouths to capture images [10, 24, 25]. However, these approaches needed time to position cats and insert sensors inside their mouths. Furthermore, special tools were needed to capture these images. Only one algorithm used camera images to determine age of white deer [2]. Side view images were used, and the deer had to be in certain position. Any slight variation in the animal's position was not acceptable. Authors used one image of white deer to measure parts of the body that represent the animal's age. Ratios of changes in body parts were measured. These ratios were then classified according to previously defined clusters. Comparisons with sets of animals with known ages were used to determine these ratios. Probability of belonging to any of the classes was then used to determine the correct age category. Many body parts such as chest and stomach depth were tested. The best average accuracy achieved in this work was 62%. This approach would give wrong results for different animal's psychology such as pregnancy. Moreover, such approach is unreliable in case of sheep because younger sheep may be larger in size than elder ones. Moreover, some sellers cheat in sheep weight by urging them to drink salted water. Correspondingly, sheep gets larger in size. Obviously, most of the previous approaches needed special tools for age estimation. They were also time consuming and cruel to animals. Therefore, the goal of this chapter is to have an automatic real time application for sheep age estimation. In this work, only one image taken by a mobile camera was needed to estimate sheep's age. Moreover, dataset of 52 teeth images for sheep of different physiological conditions and ages was used to test the proposed approach.

8.3 Theory and Background

In this chapter, some of the used methodologies in this work will be discussed briefly in the following subsections.

8.3.1 Active Contours

Active contour model is a popular computer vision method which is also known as snakes. Snakes extract contours from images by minimizing the energy regardless of the amount of noise in the image and the deformation of the objects [21]. The snakes segment objects according to the energy around are as illustrated in the following equation:

$$E_{snake} = E_{int} + E_{ext} \tag{8.1}$$

where E_{int} is the internal energy inside the object which can be defined as the square of the distance between points. While E_{ext} is the external energy which is not affected by the object. Obviously, active contours usually depend on the image intensities of the contours. However, it is prone to local minima. Therefore, the new technique of active contours without edges implemented in [29] is favored because it did not depend on edges. Moreover, this new algorithm could work on smooth images. Therefore, the initial contours could be at any place in the image. In [29], mean curve evolution was done by minimizing energy partitionly so that it was independent of the gradient. It followed the Mumford Shah function to stop the evolution. One initial curve was needed which can be anywhere in the image and it is also robust to noise. This algorithm has already been used to segment human teeth from x-ray images. The energy in this work was defined as:

$$E(contour) = \int_{inside(c)} |I_0(x,y)c_1|^2 dxdy + \int_{outside(c)} |I_0(x,y)c_2|^2 dxdy \tag{8.2}$$

where c1, c2 and I0 are intensities, cs are the mean intensities, while I_0 resembles the in and out intensities of the contour. However, this algorithm did not correctly segment most of the teeth in the experimented images. Chan et al. [6] also used active contours without edges, but they extended the scalar algorithm by minimizing the well-known classical Shah function over contour length with the error in RGB image. As a result, objects with and without edges gradient were detected. Incomplete objects were also detected. Moreover, this algorithm was robust to noise so that no preprocessing was needed.

In [18], another extension of Shah model was also suggested with multiple phases to avoid overlapping between contours.

8.3.2 Blob Detection and Counting

Blobs stand for Binary Large Objects. Blobs are the connected areas in white and black images [27]. Therefore, RGB images should be first converted into black and white images to find the connected components in a chain. From any point in the blob, neighbors are checked and tracked in all directions, north, east, west, south, northeast, southeast, southwest, northwest. Then each detected blob is given a unique label for counting. Moreover, centroids of the detected objects are determined to be able to measure a variety of blob properties.

8.3.3 Morphological Operations

Morphological operations depend on the relation between pixels and their local neighbors in contiguous pixels [12]. Erosion and dilation are well-known morphological operations that are vital in most of the visual analysis methodologies. Dilation is the process of enlarging the objects by adding pixels to the borders of the objects. On the other hand, erosion eliminates from the total mass of the object. Both erosion and dilation and all other morphological operations are dependent on the structure of given objects in the images. Erosion and dilation are often used to separate the connected components that should not be considered as one object.

8.3.4 Dentition

Dentition is the structure of teeth inside the mouth of any creature [13, 14]. Teeth development and arrangement helps in the age estimation of almost all the ruminant animals. Some of the animals have teeth in the upper jaw only, in the lower jaw only, or in both.

8.3.5 Image Collection and Camera Setting

There is no dataset available for sheep teeth. For this reason, mouth images were captured for sheep at the 6th of October farm in Ismailiya. These sheep are of the Barki breed. Egypt has a total of about 3 million sheep of three main breeds: Rahmani, Ossimi and Barki [1]. The Barki breed's origin is North Africa. However, its name is Libyan and it is found in Libya but extends to Alexandria in Egypt. Ossimi's origin is Giza in Egypt, and Ossimi is a village in Cairo. On the other hand, Rahmani's origin is Syria and Turkey. However, Rahmania is also a name of a village in Cairo. All of the three breeds have almost very similar features. All of them have fat tails with heavy wool and they are of a medium size. Samples of Barki sheep images are shown in Figure 8.1.

Selected sheep had different ages from 2 months to 5 years. Images were taken at a distance of average 0.4 m from the sheep. As illustrated in

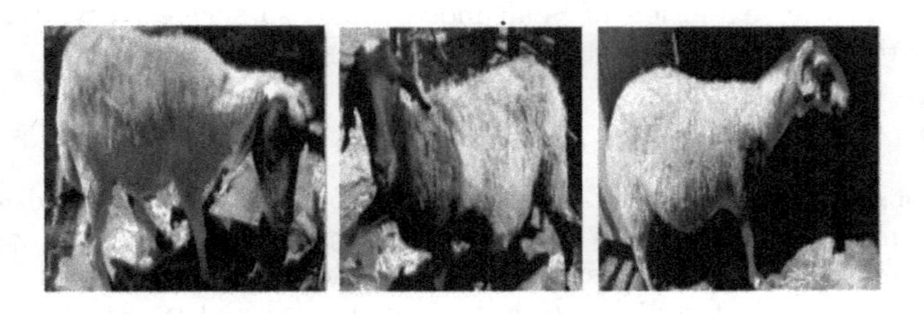

FIGURE 8.1: Samples of the selected Barki sheep.

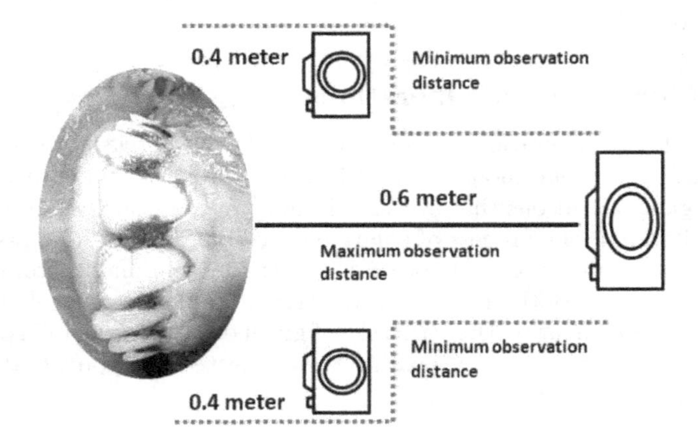

FIGURE 8.2: Camera model setting.

Figure 8.2, images were taken at a very close distance to sheep's mouth to get clear images of all sheep teeth because they affect the age calculation. Images were collected for 52 sheep. As a result, the used dataset consists of 52 teeth images to work on and test the feasibility of the proposed approach. Images were taken by a high resolution mobile camera. A professional camera was not used because the goal of the proposed approach is to be applicable on mobile phones. Sample of the collected teeth images are shown in Figure 8.3.

8.4 The Proposed Automatic Sheep Age Estimation System

The goal of this chapter is to represent a real time application to enable buyers and sellers to estimate a sheep's age. Moreover, in this chapter, age has been estimated from one single image of sheep's teeth captured by a mobile camera. Images were captured by mobile camera for sheep of different ages. Teeth images were analyzed to count and measure the linear dimensions of teeth. As a result, age can be estimated by dentition. The proposed age estimation system consists of five phases: (1) Pre-processing phase, (2) Segmentation phase, (3) Post-processing phase, (4) Age estimation phase, and (5) Validation phase. These phases are described in details in the following subsections. The overall architecture of the proposed system is also illustrated in Figure 8.4.

8.4.1 Pre-processing Phase

All images had to be processed to prepare teeth images for segmentation. Some teeth images contained flesh and bone around the teeth. Moreover, sometimes human hands had to be involved in opening sheep's mouth so they might appear in some images. Therefore, MATLAB processing tools were used to crop out these additional noisy features as shown in Figure 8.5. The images had also to be converted to grey level for segmentation using active contours. Brightness was also adjusted, as some of the images were so dark and others were very bright.

8.4.2 Segmentation Phase

Segmentation is a vital step in almost all the image processing and visual analysis algorithms. The main goal of segmentation is to decompose the image to recognize only the regions of interest. In this work, sheep teeth were segmented out of the mouth image using active contours. This approach did not need human intervention to set the initial place of contours. The contours can start at any place in the image regardless of the place of the object. Moreover, no training was needed. Active contours are based on curve evolution for object detection. The basic idea in active contours is to evolve a curve.

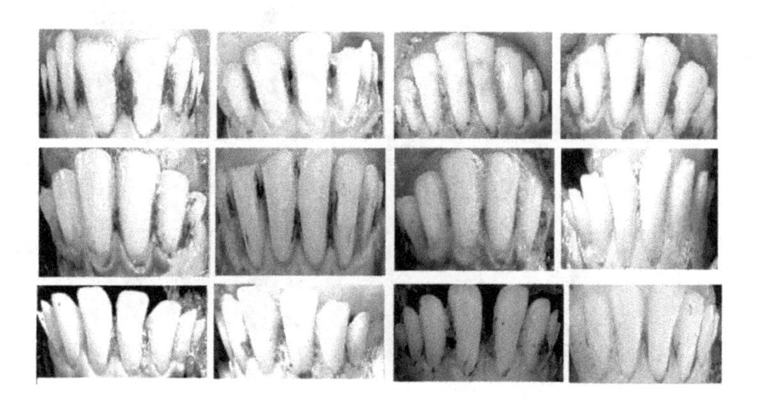

FIGURE 8.3: Teeth samples.

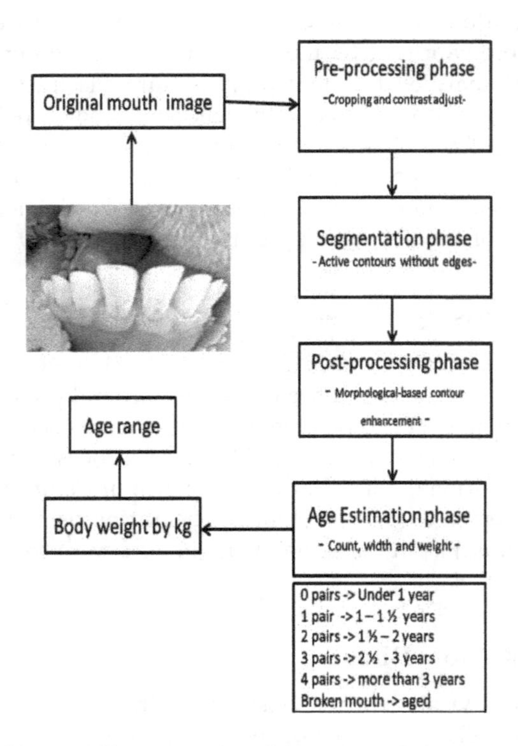

FIGURE 8.4: The general architecture of the proposed system.

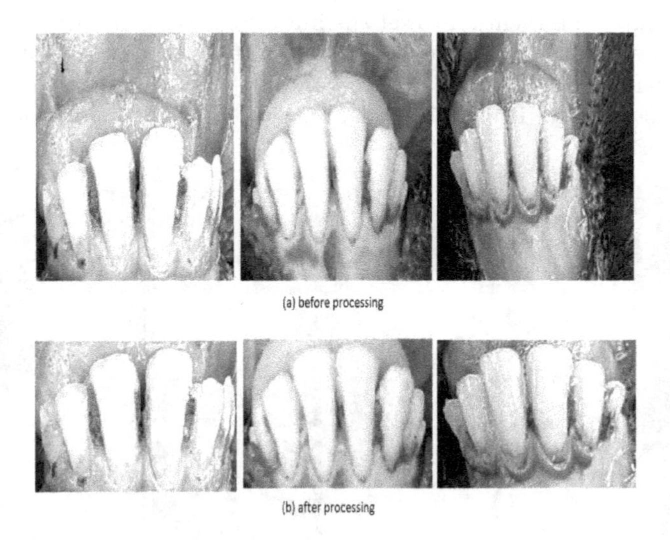

FIGURE 8.5: Images before and after processing.

A curve should start in the area that surrounds the object. Then the curve evolves until it reaches the borders of the foreground objects. However, active contours without edges can start from any place in image without presence of edges. Active contours without edges can work on smooth images and it is also robust to all types of noise. Algorithms of active contours without edges implemented in [6, 9, 18] were tested in our work as shown in Figure 8.6. As shown, the algorithm implemented in [6] gave the best results, so it was selected to be used for teeth segmentation in the proposed approach.

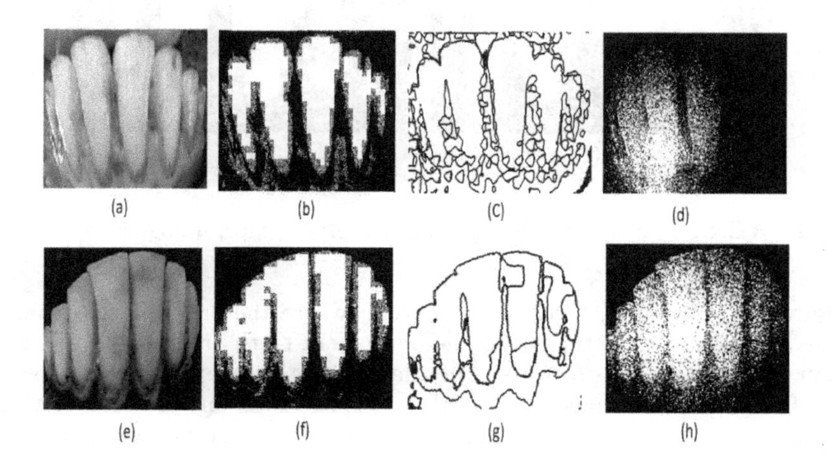

FIGURE 8.6: (a) and (e) are original images before segmentation, (b) and (f) are images after segmentation using active contours for vector-valued images in [6], (c) and (g) are segmented using multiphase active contour in [18], (d) and (h) are results of segmentation using active contour without edges implemented in [9].

8.4.3 Post-processing Phase

Morphological operations were applied to remove noise from images and to have a robust algorithm to illumination conditions as shown in Figure 8.7. Erosion and dilation were applied to eliminate undesired connectors between the extracted teeth blobs. As a result of these morphological operations, some noise might get detected as very small contours compared to other desired objects. Therefore, smaller contours were neglected and all holes were filled for noise removal.

8.4.4 Age Estimation Phase

To estimate a sheep's age, teeth size and count should be determined. Sheep have a thick pad in the upper jaw instead of teeth as was shown in

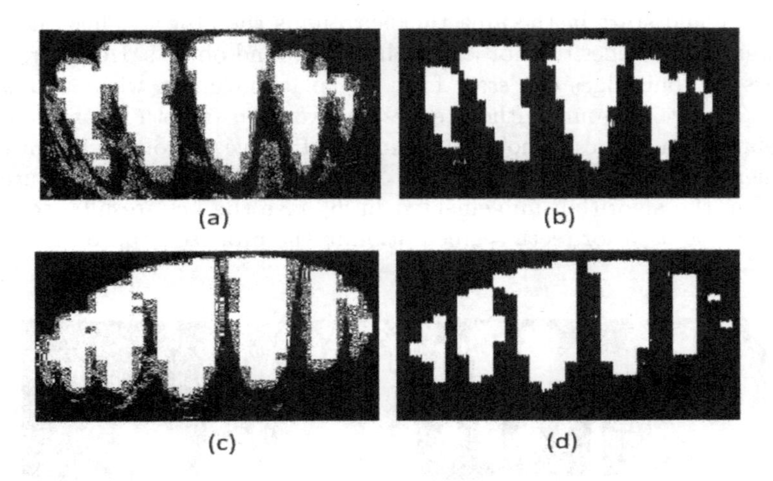

FIGURE 8.7: Segmented images after applying morphological operations.

Figure 8.5. Therefore, teeth in the lower jaw are the only features that can be extracted for age estimation. The permanent teeth are the tall ones in the lower jaw. The number of pairs of permanent teeth represents the sheep's age. If a sheep has no permanent teeth, then it is less than one year old. If a sheep has only one pair, then its age ranges from one to one and half years. If a sheep has two pairs of teeth, then its age is in the range from one and a half to two years. Moreover, if it has three pairs, then its age ranges from two and a half years to three years. Furthermore, if it has four pairs, then its age ranges from three years to four years [3, 26]. Sheep that are older than four years usually have broken teeth. However, sheep more than 4 years old are not usually kept in the farm. They are sold to be used in food production before its meat loses its tenderness and also before sheep loses its ability to cull. Age ranges are illustrated in Figure 8.8. Obviously, length and count of teeth were the best features for sheep age estimation. Teeth width was ignored in determining teeth type because aged sheep may have thin similar to milk teeth of those younger sheep. As a result, length was the only linear dimension that was considered in age calculation as shown in Figure 8.9. Blobs of extracted teeth were detected, counted and then their linear dimensions were measured. Only the maximum height of the teeth was considered. However, width was also measured for noise reduction. Segmented objects of width less than 9 pixels were considered as noise and were ignored.

It was also found that the average length of milk teeth through all the images was 21 pixels. Therefore, if two teeth had height more than 21 pixels, then they were considered as one pair to be counted for age calculation. In addition, in some cases, odd numbers of teeth were found such as 3 teeth of height more than 21 pixels. In such cases, one pair and one tooth were

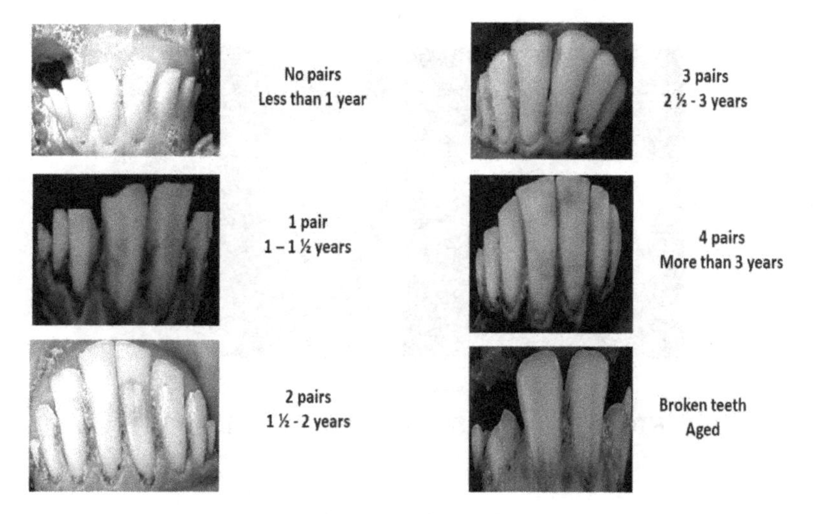

FIGURE 8.8: Sheep age ranges.

recognized as 2 pairs. Therefore, if one tooth was found higher than 21 pixels with the absence of its pair, then it was considered as another pair. Such cases are shown in Figure 8.10.

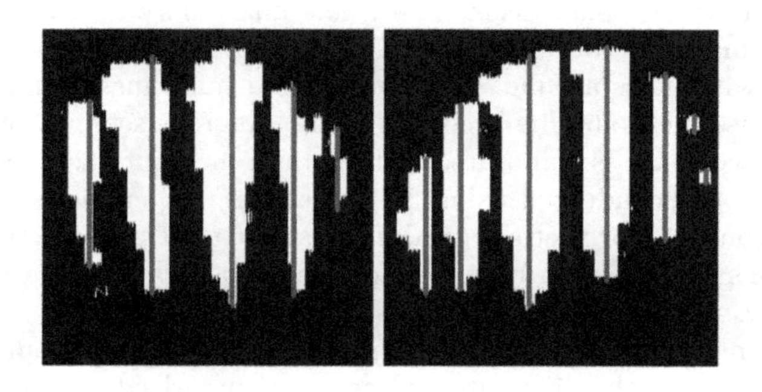

FIGURE 8.9: Teeth height.

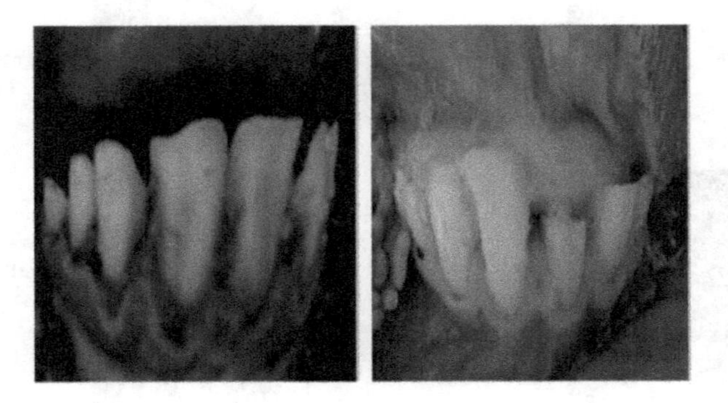

FIGURE 8.10: Special cases of teeth count.

8.5 Experimental Results and Discussion

Different active contours with different settings have been experimented. Normal Chan and Vese active contours, active contours for vector-valued images, and multiphase active contours were tested. Different settings such as number of iterations, phases and masks were also tested. The tested masks were small circular mask, medium circular mask, large circular mask, whole mask and the whole-small mask. The whole mask is a mask with holes. The whole-small mask is a two-layer mask: One layer of small circular mask and the other layer with holes around. Moreover, whole-small mask only works with the multiphase method. The accuracy of the different settings in estimating the correct age range is elaborated in the graph in Figure 8.11. These experiments have been conducted using 300 iterations for generic active contours and active contours for vector valued images. For multiphase active contours, 100 iterations were enough because incrementing iterations did not improve the accuracy.

The settings that gave the best output for each active contours methodology after testing all the different settings are summarized in Table 8.1. Incrementing the number of iterations more than the numbers given in the table did not make any difference.

TABLE 8.1: Different settings for active contours

Active contour method	Number of iterations	Mask type	Number of phases
General Chan and Vese	300	large	not applicable
Multiphase	100	whole	2
Vector	300	whole	not applicable

Active contour for vector valued images is elaborated through the different iterations in Figure 8.12. Multiphase active contour is also shown in Figure 8.13 and general Chan and Vese method in Figure 8.14.

Count and height of teeth were then calculated to determine sheep age. Accuracy of the proposed approach was measured for validation using the following equations:

$$C_{acc} = \frac{CP}{TP} * 100 \tag{8.3}$$

$$errorrate = (1 - (CP/TP)) * 100 \tag{8.4}$$

where C_{acc} is the classification accuracy, CP is the correct prediction, and TP is the total predictions. Correct predictions in this work were the accurately estimated ages in the correct age ranges. The correct predictions in this approach were 47. In other words, only 5 ages were estimated in the false age ranges. The accuracy of proposed approach in this study is 90.38%. The achieved accuracy proves the feasibility and efficiency of the proposed approach. This accuracy was achieved using active contours for vector valued images with the whole mask and for 300 iterations to extract teeth. Error rate is 9.62. However, the work done was tested on only 52 images. If more images were collected, then the accuracy might be improved because only 5 images were incorrectly analyzed. The algorithm proposed in this chapter is robust to many challenges that could be met such as dirt or flies that may cover most of the features in teeth as shown in Figure 8.15. Highest width and

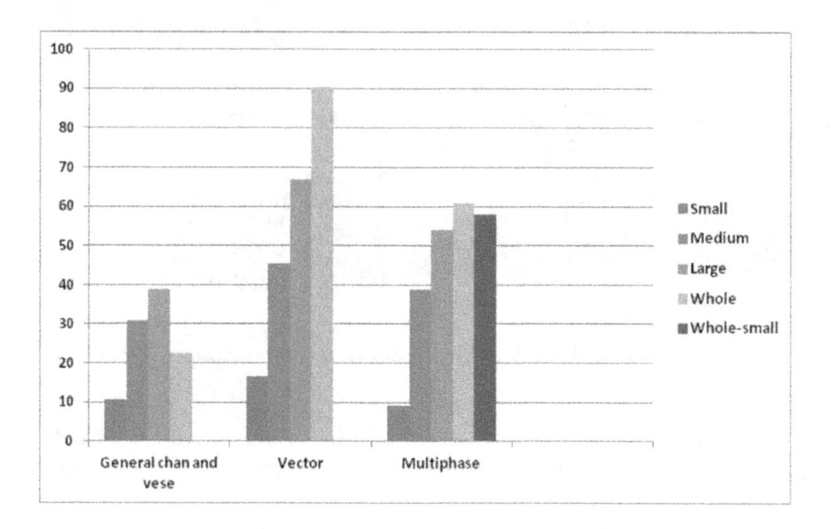

FIGURE 8.11: Accuracy results of the different active contours methodologies using different mask types.

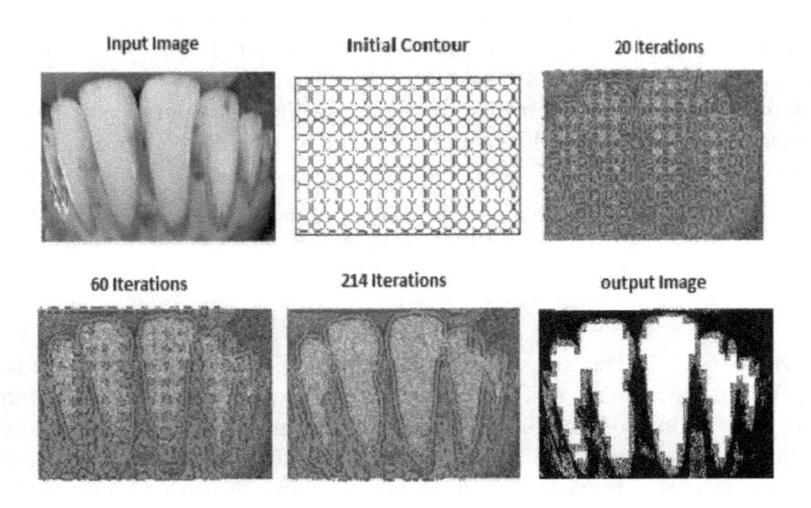

FIGURE 8.12: Active contours for vector valued images.

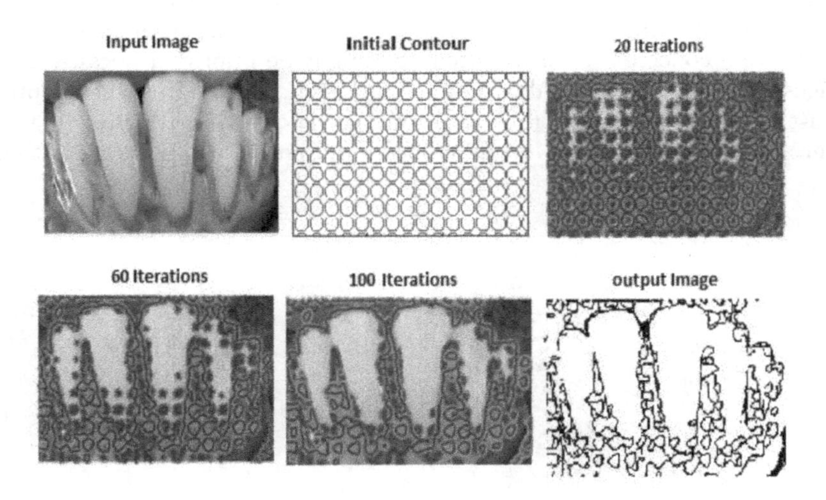

FIGURE 8.13: Multiphase active contours.

height of extracted teeth were considered as teeth height and width. For this reason, noise did not affect the results. As mentioned before, literature never tackled automatic sheep age estimation. As a result, no previous techniques were available to compare their results with the proposed approach in this chapter.

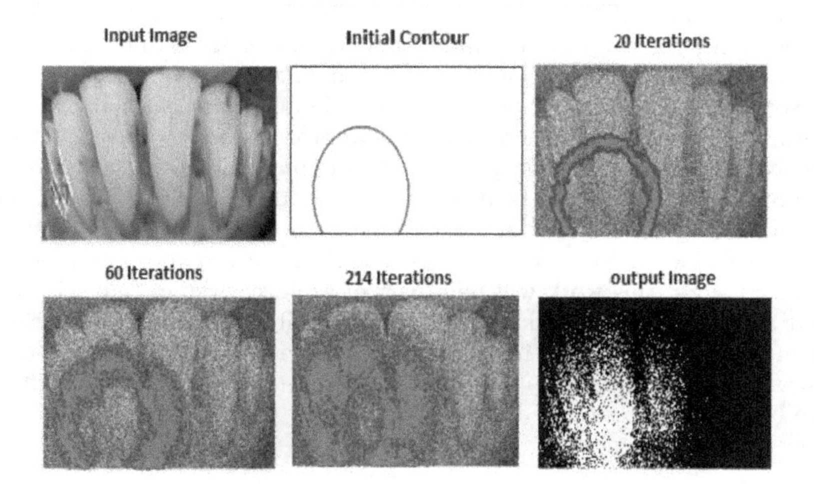

FIGURE 8.14: General Chan and Vese active contours.

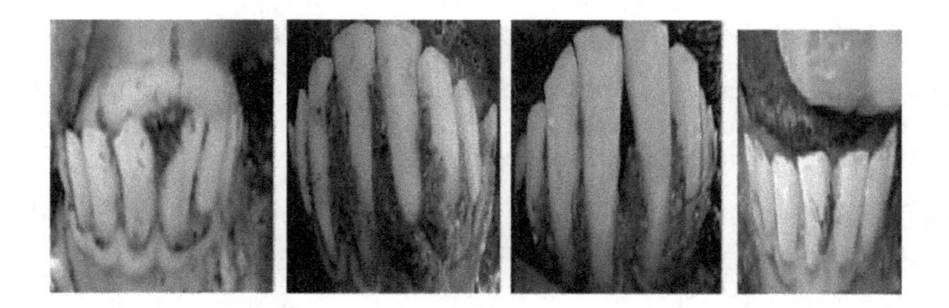

FIGURE 8.15: Images with added noise.

8.6 Conclusion and Future Work

In this chapter, a novel approach of sheep age estimation was presented. Moreover, most of the previous approaches for animal age estimation required human intervention, special environment and time-consuming techniques. In this work, sheep age estimation was completely automated in real time using images taken by mobile camera. Different approaches for teeth segmentation have been tested in this work. The proposed approach used the modern technique of active contours without edges to segment teeth from images. Then count and linear dimensions of teeth were determined to estimate sheep age using morphological operations. Accuracy of the proposed approach is 90.38%. For future work, this work will be a part of a bigger system that will be released as a mobile application which is responsible for sheep identification. Part of this system is to estimate the sheep's age because it is an important feature that helps in sheep's recognition. Moreover, more images of different sheep breeds will be collected to be added to our dataset.

Bibliography

[1] Sheep development program in egypt, available at: http://om.ciheam.org/om/pdf/c11/96605538.pdf.

[2] System and method for estimating animal age, available at: http://www.freepatentsonline.com/20160140706.pdf.

[3] *How to tell the age of sheep, Available at: https://www.dpi.nsw.gov.au/data/assets/pdf-file/0004/179797/aging-sheep.pdf.* Primefact, 2016.

[4] Aly M. Bosmans N., Ann P. and Willems G. The application of kvaal's dental age calculation technique on panoramic dental radiographs. *Forensic Science International*, 153:208–212, 2005.

[5] Priyadarshini C. and Manjunath PP. Dental age estimation methods: A review. *Int J Adv Health Sci*, 1:19–25, 2015.

[6] Sandberg BY., Chan T. and Vese L. Active contours without edges for vector-valued images. *J. of Visual Communication and Image Representation*, 11:130–141, 2000.

[7] Kuehn DW. and Berg WE. Use of radiographs to age otters. *Wildlife Society Bulletin*, 11(41):68–70, 1983.

[8] Goodwin EA. and Ballard WB. Use of tooth cementum for age-determination of gray wolves. *Journal of Wildlife Management*, 49:313–316, 1985.

[9] Chan T. F. and Vese L. A. Active contours without edges. *IEEE: Transactions on Image Processing*, 2:266–277, 2001.

[10] Knowlton FF. and Whittemore SL. Pulp cavity-tooth width ratios from known-age and wild-caught coyotes determined by radiography. 29:239–2441, 2001.

[11] Tobias G., Tobias TA. and Abood SK. Estimating age in dogs and cats using ocular lens examination. *Comp. Cont. Educ. Pract.* 22:1085–1091, 2000.

[12] Gil J. and Kimmel R. Efficient dilation, erosion, opening and closing algorithms. *IEEE: Transactions on Pattern Analysis and Machine Intelligence*, 24:1606–1617, 2002.

[13] Rackham J. A comparison of methods of age determination from the mandibular dentition of an archaeological sample of cattle. *Teeth and Anthropology, BAR British Series*, 291:149–168, 1986.

[14] Pace JE. and Wakeman DL. Determining the age of cattle by their teeth. *CIR253, Florida Cooperative Extension Service, Institute of Food and Agricultural Sciences*, University of Florida, Gainesville, FL, 2003.

[15] Whittemore SL. and Knowlton FF. Pulp cavity-tooth width ratios from known-age and wild-caught coyotes determined by radiography. *Wildlife Society Bulletin*, 29:239–244, 2001.

[16] Thomsen IO., Kvaal SI., Kolltveit KM. and Solheim T. Age estimation of adults from dental radiographs. *Forensic Science International*, 74(41):175–185, 1995.

[17] Thomsen IO., Kvaal SI., Kolltveit KM. and Solheim T. Age estimation of adults from dental radiographs. *Forensic Science International*, 74:175–185, 1998.

[18] Vese L. and Chan T. A multiphase level set framework for image segmentation using the mumford and shah model. *Int'l J. Computer Vision*, 50:271–293, 2002.

[19] Peterso RO. Landon DB., Waite CA. and Mech LD. Evaluation of age determination techniques for gray wolves. *Journal of Wildlife Management*, 62:674–682, 1998.

[20] Dix LM. and Strickland MA. Use of tooth radiographs to classify martens by sex and age. *Wildlife Society Bulletin*, 14(41):275–279, 1986.

[21] Acharya R. Luo H., Qiang L. and Gaborski R. Robust snake model. *IEEE Conf. on CVPR*, pages 452–457, 2000.

[22] del Campo M. San Julin R. Montossi F., Furnols M. and C. Saudo. Sustainable sheep production and consumer preference trends: compatibilities, contradictions, and unresolved dilemmas. *Meat Sci*, 9:772–789, 2013.

[23] Nielsen HN., Víkingsson GA., Hansen SH., Ditlevsen S. and Heide-Jørgensen MP. Two techniques of age estimation in cetaceans: Glgs in teeth and earplugs, and measuring the aar rate in eye lens nucleus. *NAMMCO Scientific Publications*, 10:1–15, 2017.

[24] Park K., Ahn J., Kang S., Lee E., Kim S., Park S., Noh H. and Seo K. Determining the age of cats by pulp cavity/tooth width ratio using dental radiography. *Journal of Veterinary Science*, 15:557–561, 2014.

[25] Tumlison R. and McDaniel VR. Gray fox age classification by canine tooth pulp cavity radiographs. *The Journal of Wildlife Management*, 48:228–230, 1984.

[26] Begaz S. and Awgichew K. Estimation of weight and age of sheep and goats. *Ethiopia Sheep and Goat Productivity Improvement Program (ESGPIP)*, 12(23), 2009.

[27] Hinz S. Fast and subpixel precise blob detection and attribution. *IEEE International Conference on Image Processing (ICIP)*, 3:450–457, 2005.

[28] Linhart SB. and Knowlton FF. Determining age of coyotes by tooth cementum layers. *Journal of Wildlife Management*, 31:362–365, 1967.

[29] Shah S., Abaza A., Ross A. and Ammar H. Automatic tooth segmentation using active contour without edges. *IEEE Biometrics Symposium*, page 1–6, 2006.

[30] Anderson JL. Smuts GL. and Austin JC. Age-determination of african lion (panthera-leo). *Journal of Zoology*, 185:115–146, 1978.

Chapter 9

Diversity Matrix Based Performance Improvement for Ensemble Learning Approach

Rajdeep Chatterjee

KIIT, Bhuvaneswar, India

Siddhartha Chatterjee

DSMS Group of Institution, India

Ankita Datta

KIIT, Bhuvaneswar, India

Debarshi Kumar Sanyal

KIIT, Bhuvaneswar, India

9.1 Introduction 196
9.2 Related Work 196
9.3 Theoretical Background 197
9.3.1 Wavelet Based Energy and Entropy 197
9.3.2 Ensemble Classification 199
9.3.2.1 Bagging ensemble learning 199
9.3.2.2 Majority voting 200
9.3.3 Used Diversity Techniques 200
9.3.3.1 Cosine dissimilarity 201
9.3.3.2 Gaussian dissimilarity 202
9.3.3.3 Kullback-Leibler divergence 202
9.3.3.4 Euclidean distance 202
9.4 Proposed Method 202
9.5 Results and Discussion 205
9.5.1 Preparing the Used Datasets 205
9.5.2 Experimental Set-up 207
9.5.3 Results Analysis 208
9.6 Conclusion and Future Work 211
Bibliography 213

9.1 Introduction

Performance enhancement of classification accuracy is a desired aspect in machine learning. It is shown in the literature that different hyper-parameter tuning is essential to obtaining high performance with classifiers. However, it is also true that dataset preparation, preprocessing, and classification algorithm influence the overall performance. Classifiers are categorized as either linear or non-linear classifiers. Naive Bayes (NB), Support Vector Machine (SVM), K-Nearest Neighbor (KNN), and Decision Tree (DT) are a few types of widely used classifiers [15, 20]. However, another category of classifier has also become popular in recent times; they are ensemble classifiers. In the following sections, we discuss the detailed concept of ensemble classification, particularly, a special type of bagging ensemble classifier called mix-bagging. Here, multiple predictors are used to build models from different bags (which are subsets of the training set) [11, 34].

The current book, *Hybrid Computational Intelligence: Research and Applications*, aims at research related to different intelligent techniques applied in the field of engineering. It includes contemporary topics such as hybridization of different intelligent computational techniques and their application to domains such as Image Processing, Signal Processing, Bio-informatics, Biomedical Engineering, etc. We propose a diversity matrix based pruning technique which uses the same outputs as those of the original predictors and select a subset of predictors for performance improvement in terms of classification accuracy. We apply it to electroencephalography (EEG) based motor-imagery left-right hand movement bio-signal classification which comes under the domain called Brain Computer Interface (BCI) [4, 35]. Thus, this chapter provides a novel way of solving the performance deficit in the EEG based bio-signal classification with a satisfactory performance in its first adaptation. Therefore, we believe the scope and relevance of this chapter is well suited to the book and in sync with its core spirit.

The chapter is organized into six sections. Section 9.2 explains related work on the EEG signal classification. Section 9.3 provides the necessary theoretical understanding for our study. The proposed diversity matrix based pruning technique is explained with multiple examples in Section 9.4. The obtained results are discussed and analyzed in Section 9.5. Finally, the chapter concludes with Section 9.6.

9.2 Related Work

Tuning of hyper-parameters of a classifier is one of the most common methods to obtain a better performance in classification. Another technique

TABLE 9.1: Different classification techniques used in motor-imagery EEG signal analysis.

References	Used Feature-sets	Classifiers
[4]	Wavelet Energy-entropy + RMS + Band Power + PSD Average Power + Statistical based	Multi Layered Perceptron (MLP) (Kernel: Radial basis function)
[7]	Wavelet energy & entropy	Support Vector Machine (SVM) (Kernel: Linear)
[9]	Adaptive Autoregressive parameters (AAR)	Extreme Learning Machine (ELM) (Kernel: Sigmoidal)
[6]	Adaptive Autoregressive parameters (AAR)	Support Vector Machine (SVM) (Kernel: Polynomial, Degree: 3)
[11]	Band Power	Ensemble learning (Adabbost: Adaptive boosting)
[10]	AAR + Band Power + Energy -entropy	Ensemble Architecture (Using different datasets but single learner type, K-Nearest Neighbor)
[2]	PSD based Average Band Power	K-Nearest Neighbor (KNN)

is use of feature selection or reduction in the data pre-processing step. Principal component analysis (PCA) [33], mutual information (MI)-based feature ranking [21, 32], and rough set theory (RST)-based approaches [19, 26, 27] are used and implemented widely. In BCI, it is important to identify the most discriminant features in EEG signals generated by various brain activities. Feeding only the features containing the most informative features increases the accuracy of a classifier. Sometimes the use of all the features may not provide a good quality classification compared to a selected subset of features.

In [18], the authors introduced rough-set theory based upper and lower approximation to calculate the γ value for dimensionality reduction in EEG data. Specifically, in the field of motor-imagery classification, researchers in [9] used the discernibility matrix to find the possible reduct (subset of features) and demonstrated classification improvement through an empirical study. The discernibility matrix always works on discretized dataset. The transformation of continuous features into discrete values incurs information loss. In [6], the authors proposed a fuzzy discernibility matrix to avoid the discretization process and obtained a very positive outcome on the motor-imagery datasets. In Table 9.1, a summary of different motor-imagery EEG signal classification techniques is given.

9.3 Theoretical Background

9.3.1 Wavelet Based Energy and Entropy

Most commonly used signal processing techniques such as Fast Fourier Transform and Short Time Fourier Transform represent signals poorly localized in time or frequency. On the other hand, wavelet localizes the input signal better in time or frequency. The wavelet transform seeks to achieve the best trade-off between temporal and frequency resolution. Instead of using sines and cosines, wavelet transform uses finite basis functions called wavelets. These single finite-length waveforms are known as the mother wavelets. The wavelet transform basically represents the original input signal in the form of a linear combination of basis functions. It comes with different shapes and sizes. From our past research work, we conclude that wavelet based energy-entropy method is well suited for EEG based motor-imagery classification [4].

Wavelet is a rapidly decaying wavelike oscillation that has zero-mean, unlike a sinusoid which extends to infinity. Discrete Wavelet Transform (DWT) has been used in this chapter for feature extraction. DWT splits the input signal in the low-pass sub-band and the high-pass sub-band. The output of these sub-bands are also known as approximation coefficients $A1$ and detail coefficients $D1$, respectively. One can further decompose the approximation sub-band at multiple levels or scales. It is derived repetitively for again two times to its 3rd level (shown in Fig. 9.1). The Daubechies (db) basis function with filter size 4 and third level detail coefficients $D3$ are used to extract features from the input EEG signal [5]. DWT is ideal for downsizing the actual input signal while retaining the properties of the original signal with fewer coefficients. It helps in the overall computing process. However, we have performed another step to reduce the dimension of the obtained wavelet coefficients by implementing Eqs. 9.1 and 9.2 [4, 5, 14, 16].

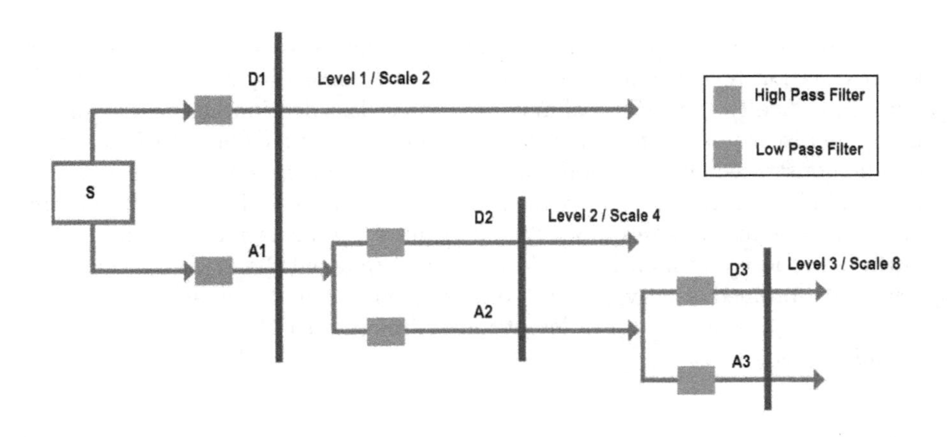

FIGURE 9.1: Third level of Discrete Wavelet Transformation into approximation and detail coefficients.

$$ENG_l^{C3} = \sum_{(n=1)}^{2^{s-l}-1} \mid C_x(l,n) \mid^2 \tag{9.1}$$

$$ENT_l^{C3} = -\sum_{(n=1)}^{2^{s-l}-1} \mid C_x(l,n) \mid^2 \log(\mid C_x(l,n) \mid^2) \tag{9.2}$$

$$N = 2^S, \qquad 1 < l < S$$

The ENG_l^{C3} and ENT_l^{C3} are the computed energy and entropy features for the lth trial of $C3$ electrode. Similarly, energy and entropy for $C4$ electrode are also obtained. $C_x(l,n)$ indicates the nth sample of lth trial as input signal. The length of the feature set used in this chapter is 8 extracted from alpha $(8-13Hz)$ and beta $(13-25Hz)$ bands using Eqs. 9.1 and 9.2.

9.3.2 Ensemble Classification

Ensemble learning is basically based on a combined decision over multiple learners. There are some real-world simple examples which can explain the concept of ensemble learning. We normally consult different doctors for their advices before going for a major surgery. Another usual example is that we purchase an electronic product after reading multiple user reviews. Our aim is to improve the confidence level before making the right decision, and ensemble learning does the same. Ensemble learning can be divided into two types: (i) independent, and (ii) dependent. In independent ensemble learning approach, the base classifiers are built independently and the results of the base classifiers are combined using suitable techniques such as voting (depicted in Fig. 9.2). In dependent ensemble learning approach, construction of next classifier depends on the output of the previous classifier [30, 31].

An ensemble approach is said to be efficient if there is *diversity* among base classifiers, i.e., its individual learners commit errors on different instances. Diversity allows an ensemble classifier to exhibit higher accuracy than the average accuracy of its individual models. Different strategies can be used to bring diversity in the ensemble model: (a) using different hyper-parameters for a single learner type, (b) using different types of classifiers (multiple types of learners), and (c) using different types of feature sets to build the prediction model [25, 36].

9.3.2.1 Bagging ensemble learning

A highly popular independent ensemble classifier method is *bagging.* According to the theory of bagging, subsets are created by sampling with replacement from the training set. "With replacement" means that some of the

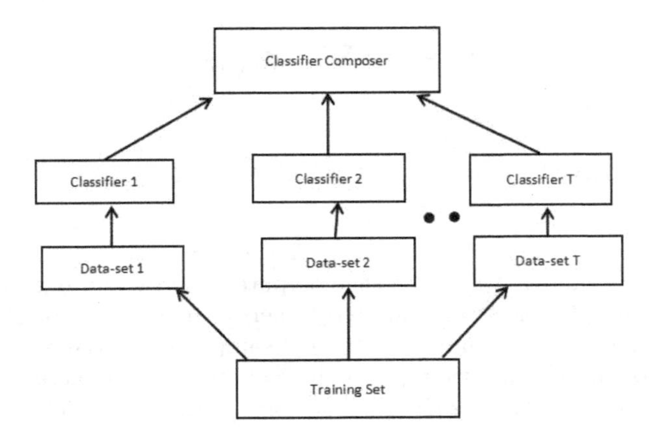

FIGURE 9.2: Diagram of an independent ensemble learning.

original instances may appear more than once in a bag. The size of the bag is less than the size of the actual training set. Multiple prediction models are built from different bags and those models are used to classify unknown instances (from test set). The prediction outputs from all the bags are combined (usually majority voting) and the final output is determined. The composite classifier (combined model of bagged classifiers) outperforms a single model for the same training set [3, 8, 29]. The pseudo-code representation of modified mix-bagging ensemble learning approach is given in Algorithm 1. It is self-explanatory.

9.3.2.2 Majority voting

It is already mentioned in the last subsection that a popular technique for combining results from multiple predictors is majority voting. It works on the simple principle of voting based on maximum similarity as shown in Figure 9.3.

9.3.3 Used Diversity Techniques

In this chapter, we have used four different types of dissimilarity measures (DM) based on: (a) Cosine similarity, (b) Gaussian similarity, (c) Kullback-Leibler divergence, and (d) Euclidean distance. Assume $\vec{X}$ and $\vec{Y}$ are two non-zero vectors of equal length. The brief description of each type is given as follows:

Algorithm 1 Modified Bagging (Mix-bagging)

1: $Inputs \quad : \quad Given(x_1, y_1), ..., (x_w, y_w) \ , where \ w \ total \ training - set \ instances$
   $x_i \ is \ a \ feature - set \ \& \ y_i \ is \ the \ corresponding \ decision - class$
   $xtest \ is \ testing - set \ of \ size \ v \ instances$
2: $-choose : m_{bag}, \ number \ of \ bags$
3: $-choose : m_{use}, \ size \ of \ bootstrap \ samples$
4: $-choose : iter, \ number \ of \ independent \ executions$
5: $-initialize : prediction^{best}_{final} \leftarrow 0$
6: **for** $r \leftarrow 1 \quad to \quad iter$ **do**
7:   **for** $t \leftarrow 1 \quad to \quad m_{bag}$ **do**
8:     $indx, \ bootstrap \ sample \ instances \ (m_{use} < w)$
9:     $X_t \leftarrow \forall \ x_i \ where \ i \ \in \ indx$
10:     $Y_t \leftarrow \forall \ y_i \ where \ i \ \in \ indx$
11:     $model_t \leftarrow learner(X_t, Y_t)$
12:   **end for**
13:   **for** $t \leftarrow 1 \quad to \quad m_{bag}$ **do**
14:     $predict_{nt} \leftarrow learner(xtest, model_t)$
15:   **end for**
16:   $prediction^{r}_{final} \leftarrow majority-voting(predict_t), \ for \ v \ test \ instances$
17:   **if** $if \quad prediction^{r}_{final} \quad > \quad prediction^{best}_{final}$ **then**
18:     $Remember \ the \ bags \ and \ store \ prediction^{best}_{final} \leftarrow prediction^{r}_{final}$
19:   **end if**
20: **end for**
21: **return** $prediction^{best}_{final}, \ for \ iter \ independent \ runs$

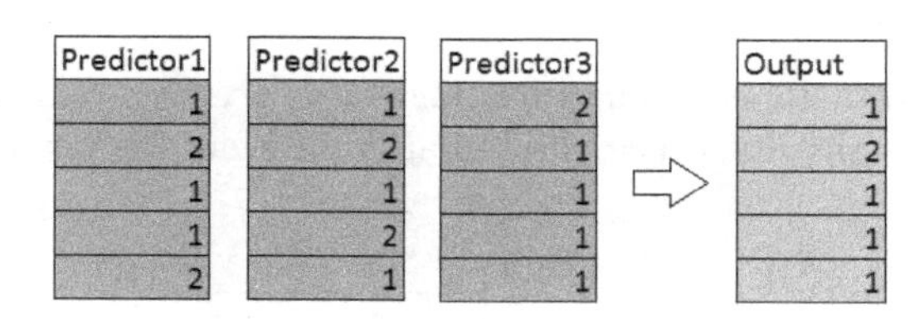

FIGURE 9.3: Example of majority voting with three predicted output vectors.

9.3.3.1 Cosine dissimilarity

Cosine similarity measures the similarity between two non-zero vectors. Although it does not respect the triangular inequality, it is a widely used similarity measure. Its value lies in the range $[0, 1]$. In our case, we need diversity (dissimilarity between vectors). In order to get that, we have subtracted the cosine similarity from 1 (shown in Eq. 9.3) [17].

$$DM1(\vec{X},\vec{Y},\sigma) = \left[1 - \frac{\vec{X}.\vec{Y}}{\|\vec{X}\|.\|\vec{Y}\|}\right] \tag{9.3}$$

9.3.3.2 Gaussian dissimilarity

The Gaussian kernel function measures similarity (distance) and its value ranges between zero and one. Many researchers prefer it over Euclidean distance due to its bounded property [0, 1]. Like cosine dissimilarity, in Eq. 9.4 we need to subtract the value from 1 to obtain the Gaussian dissimilarity [12, 23] .

$$DM2(\vec{X},\vec{Y},\sigma) = \left[1 - e^{-\frac{\|\vec{X}-\vec{Y}\|^2}{2\sigma^2}}\right] \tag{9.4}$$

9.3.3.3 Kullback-Leibler divergence

Kullback-Leibler divergence (KLD), which is also called relative entropy, is related to information divergence and information for discrimination. It measures the difference between two probability distributions [12, 22]. It is presented in Eq. 9.5 for discrete scenario.

$$DM3(\vec{X},\vec{Y},\sigma) = \left[\sum_i \vec{X}_i \log \frac{\vec{X}_i}{\vec{Y}_i}\right] \tag{9.5}$$

9.3.3.4 Euclidean distance

The Euclidean distance, sometimes known as Pythagorean metric, is the most popular diversity (distance/dissimilarity) measure (shown in Eq. 9.6).

$$DM4(\vec{X},\vec{Y},\sigma) = \left[\sqrt{\sum_i (\vec{X}_i - \vec{Y}_i)^2}\right] \tag{9.6}$$

9.4 Proposed Method

The prime focus of the proposed method is to improve the already obtained classification accuracy from an ensemble by majority voting. Assume an ensemble classifier with five learners (with five different bags) which predicts decision class 1 or 2 for three test cases (refer to Tables 9.2, 9.5 and 9.6). The corresponding sixth and seventh columns indicate the actual class labels and combined results after majority voting. Let us discuss the proposed technique for Table 9.1. The accuracy of this example is $\frac{2}{4}$ (i.e., acc1=50%). As there are five learners; therefore, a diversity matrix is formed based on the diversities obtained from each pair of predictors (the results of 5 learners built from 5 bags). The diversity matrix shown in Table 9.3 is a symmetric matrix. Next, we need to identify the top three diversity values from Table 9.3. Subsequently, we also calculate the total diversity contributed by each predictor based on the top diversity values only.

TABLE 9.2: Assumed predictors-set-I.

P1	P2	P3	P4	P5	Actual	Voted
2	2	1	2	1	1	2
1	2	2	2	1	1	2
1	2	2	2	2	2	2
1	1	2	1	1	1	1

TABLE 9.3: Diversity matrix of Table 9.1 using cosine dissimilarity.

	P1	P2	P3	P4	P5
P1	0	.0565	.1614	.0565	.1429
P2	.0565	0	.0769	0	.0565
P3	.1614	.0769	0	.0769	.0565
P4	.0565	0	.0769	0	.0565
P5	.1429	.0565	.0565	.0565	0

The top three diversity values are 0.1614, 0.1429 and 0.0769. The sum of diversities generated by each predictor for Table 9.3 are:

$$s\vec{M} = 0.3042, 0.0769, 0.3152, 0.769, 0.1429$$

Now, based on the $s\vec{M}$, the top three predictors are selected and the rest are discarded. This process in our proposed method is called *pruning*. The primary objective of this step is to use fewer predictors in ensemble with the expectation to improve the overall classification performance. Here, the top three predictors are 3, 1 and 5. The pruned table is displayed in Table 9.4.

From Table 9.3, we have the improved accuracy of $\frac{4}{4}$ (i.e., acc2=100%). Therefore, the improvement in performance from the same set of predictors is

TABLE 9.4: Pruned table obtained from Table 9.2.

P3	P1	P5	Actual	Voting
1	2	1	1	1
2	1	1	1	1
2	1	2	2	2
2	1	1	1	1

increased by 50% with respect to the original result. It must be noted that the value **three** is selected as pruned number of predictors because it is the only possible odd number below five which is the actual number of learners used in the example. We have used two other examples to examine the proposed method further. The assumed cases are shown in Tables 9.5 and 9.6.

TABLE 9.5: Assumed predictors-set-II

P1	P2	P3	P4	P5	Actual	Voting
1	2	1	2	1	1	1
1	1	1	1	2	1	1
2	1	1	2	2	2	2
2	2	2	2	2	1	2

TABLE 9.6: Assumed predictors-set-III

P1	P2	P3	P4	P5	Actual	Voting
1	2	2	2	1	1	2
1	1	1	2	2	1	1
1	2	2	1	1	2	1
1	1	1	1	1	1	1

The analysis of last two cases (Tables 9.5 and 9.6) is summarized and presented in Table 9.7. It is observed that the improvements in performance for Tables 9.5 and 9.6 are 0% and 25%, respectively. The proposed method is also given as pseudo-code in Algorithm 2.

TABLE 9.7: Different performance values obtained for Tables 9.5 and 9.6 using proposed algorithm (2).

Top-3 Diversity Values	Total Diversities per predictors	Top-3 Predictors	Original Accuracy (acc1 %)	Accuracy after pruning (acc2 %)
0.1229 0.1000 0.0769	0.1000 0.2229 0.0000 0.0769 0.1999	2 5 1	75	75
0.1633 0.1000 0.0551	0.0551 0.2633 0.2633 0.2000 0.3818	5 2 3	50	75

9.5 Results and Discussion

9.5.1 Preparing the Used Datasets

The widely used Fisher's Iris dataset or simply Iris data from UCI Machine Learning Repository [13] is used prior to the implementation of proposed algorithm on our desired motor-imagery EEG data to re-validate our claim. The Iris data has 4 features (sepal and petal lengths and widths) and 3 different class labels (*Iris setosa*, *Iris virginica* and *Iris versicolor*) with 50 instances of each type. In order to maintain the uniformity in this chapter, we have used any two combination of classes from Iris to reformulate it as a binary classification problem before applying the proposed performance improvement technique.

The BCI competition-II Dataset-III from the Department of Medical Informatics, Institute for Biomedical Engineering, University of Technology (Graz) has been used in our study [1, 28]. There are a total of 280 trials (instances). The decision classes are 1 and 2 which signify the left-hand movement and the right-hand movement, respectively. It has 140 trials for left-hand movements and the remaining 140 trials are for right-hand movement. Again, out of a total of 280 trials, the first 140 instances have been recommended for training and the remaining 140 instances are kept for testing purposes. The dataset consists of three EEG channels: $C3$, Cz and $C4$. We have considered only the $C3$ and $C4$ electrodes because the motor-cortex regions of the brain under these electrodes are dominant in actual limb (hand) movement or even in imagination of such movement. The length of actual motor-imagery EEG signal of our interest is of 6 seconds (Figure 9.4). The brief details of all the

Algorithm 2 Diversity Matrix Based Pruning (proposed algorithm)

1: *Inputs : Given* $P_1 ... P_m$ *, where m is the total number of predictors*
  $\vec{A}$ *is the actual decision class vector*
  $\vec{A'}$ *is the original predicted decision class vector (obtained before pruning)*
2: *Create Diversity Matrix* **cM** *using different diversity measures* DM^* (1. *cosine*, 2. *gaussian*, 3. $KL-divergence$ *and* 4. *eulcedian distance*)
3: *Select the pruned k numbers of predictors* $\vec{P_i}$
  (*in our research, it is* $\lceil \frac{m}{2} \rceil$)
4: *Find the top k highest diversity values* $top\vec{D}_k$ *from* **cM**
5: *Compute the total diversity (values)* $s\vec{M}$ *contributed by each predictors (i.e.* P_i *based on* $top\vec{D}_k$)
6: *Sort the* $s\vec{M}$ *in descending order and identify the top k diversity contributors*
7: *Create a* $k-subset$ *of predictors*
  (*prune the predictors contributed less diversity*)
8: *Apply majority voting to obtain the new decision class vector* $\vec{A''}$
9: *Using* $\vec{A}$ *and* $\vec{A'}$, *calculate the original accuracy* ($acc1$) *and similarly using* $\vec{A}$ & $\vec{A''}$, *calculate the new accuracy* ($acc2$)
10: **if** $acc2 > acc1$ **then**
11:   *keep the improved result i.e.* $acc2$
    $ACC = acc2$
12: **else if** $acc2 == acc1$ **then**
13:   *Select either of the solution* ($acc1$ *or* $acc2$)
    $ACC = acc1$ *or* $acc2$
14: **else**
15:   *Retain the original accuracy obtained from* $\vec{A'}$
    $ACC = acc1$
16: **end if**
17: **return** *Accuracy* (ACC)

used datasets are given in Table 9.8.

TABLE 9.8: Description of used datasets.

Datasets	Acronyms	Descriptions
Iris	DS-I	Class 1 & 2 (binary classification)
Iris	DS-II	Class 1 & 3 (binary classification)
Iris	DS-III	Class 2 & 3 (binary classification)
BCI competition II Dataset III	DS-IV	Motor-imagery EEG signal (binary classification)

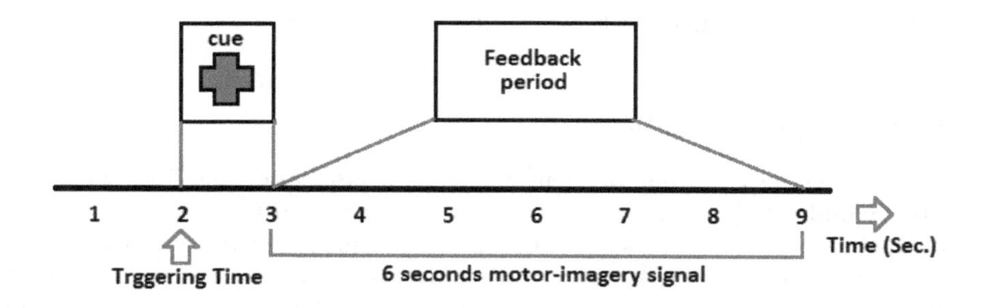

FIGURE 9.4: Motor-imagery EEG signal of 6 seconds for each trial.

9.5.2 Experimental Set-up

A computer with an Intel Core $i5$ processor, 8GB RAM and Windows 10 Pro operating systems has been used. MATLAB $R2016a$ is used for feature extraction, selection and classification. The important aspects of our experimental set-up are outlined below.

1. *Iris setosa*, *Iris virginica* and *Iris versicolor* of the Iris dataset are mapped to class labels 1, 2 and 3, respectively.
2. It is already discussed in an earlier subsection, the Iris dataset is used as three separate binary classification problems rather than a single multi-class (more than two classes) classification problem.
3. The datasets DS-I, DS-II and DS-III are formed from Iris classes 1 & 2, Iris classes 1 & 3 and Iris classes 2 & 3 combinations (total instances: 100, training: 60% and testing: 40%).
4. Then, the raw EEG signal is filtered between 0.05Hz and 50Hz using an elliptic band-pass filter with a 128Hz sampling rate.
5. The fourth dataset DS-IV is formed using wavelet based energy-entropy feature extraction (for alpha $8 - 13Hz$ and beta $13 - 25Hz$ bands only) technique used on the filtered signal for each electrode $C3$ & $C4$.
6. The mix-bagging ensemble learning is applied for classification with majority voting on all the four datasets (DS-I, DS-II, DS-III & DS-IV).
7. Proposed Diversity Matrix based pruning algorithm is implemented (with four different diversity measures discussed in Section 9.3.3) on the obtained predictors.

8. Finally, the k ($< m$) number of predictors are selected based on Algorithm 2 and calculate the new classification accuracy.

9.5.3 Results Analysis

Mix-bagging is an independent ensemble classification method. The concept behind mix-bagging is to use multiple learners instead of a single type of classifier (shown in Table 9.9). It is proved from the available research that diversity in an ensemble brings overall improvement in performance. However, this chapter aims at a very different aspect of performance improvement in the classification problem. The proposed pruning algorithm is an additional method used at the end of the procedure to further improve the classification accuracy. In our three examples (Tables 9.2, 9.5 and 9.6), the performance has either improved or remained the same. However, if we apply the proposed method for performance enhancement to the mentioned datasets, sometimes performance also degrades relative to the original result (obtained in the normal course without using the proposed method). Performance degradation is due to the similar results obtained from the different bags when the total number of instances is high (each bag uses a random set of instances of the same size from the actual training set, i.e., m_{use}). In simple words, the diversity among the results obtained from each bag is insignificant. Therefore, two possible solutions can be proposed: (i) the used ensemble classifier should generate more diverse results; or/and (ii) the diversity measure needs to discriminate among the different predictors even when they have low dissimilarity among them. In order to examine the reliability of the proposed method, the algorithm is implemented with different types of datasets with variations in terms of m_{use} and diversity measures. The proposed method is executed 100 times with 5 independent runs and the following measures are taken as results in Tables 9.10, 9.11, 9.12 and 9.13:

1. **Diversity Measure:** Type of diversity measure used in the proposed algorithm to develop the diversity matrix.

2. **Degraded:** The percentage of degradation in performance after application of the proposed method.

3. **Improved:** The percentage of improvement in performance after application of the proposed method.

4. **Mean (%) Improved:** The mean of overall improvements in 5×100 runs for a specific diversity measure.

5. **Mean #Mo5:** The mean of maximum improvements for 5 independent runs (#Mo5).

6. **#Mo5:** The maximum improvement out of 5 independent executions.

The classification accuracy (acc1) obtained from both the DS-I and DS-II datasets using the mix-bagging classifier is 100%. Therefore, no further improvement is possible. However, results obtained from the dataset DS-III is below 100% and so, our proposed pruning algorithm is suited for exploring performance enhancement. The detailed observations are available in Table 9.10. As expected from the examples, the results are positive and encouraging. Here, m_{use} is kept fixed at 45, i.e., 75% of the total 60 training instances. The best mean percent improvement and #Mo5 are 2.84% and 7.50%, respectively. The highest accuracy after applying the proposed pruning technique is 97.50% which is an improvement of 7.50% from the original accuracy (acc1) of 90%. Out of all four diversity-based pruning techniques for performance improvement, the Kullback-Leibler Divergence (KLD) method provides the most improved (in percentage %) cases from 5 independent runs of 100 times.

The empirical results are quite fascinating for dataset DS-IV. It is observed that the mean percent improvement and mean of #Mo5 are more or less similar for different values of m_{use} (used bag sizes are 80, 100 and 120). The results in Tables 9.11, 9.12 and 9.13 suggest the stability of the proposed method irrespective of the quality of the predictors (built with different bag sizes). The other significant observation is that with the increase in the bag size, i.e., m_{use} value, the overall percentage of improved cases increase over degradation. However, irrespective of bag sizes, the Gaussian diversity measures always provide a higher percentage of improvement cases than degradation cases. The highest mean percent improvement and mean of #Mo5 are 1.75% and 5.28%, respectively, after using our proposed pruning algorithm. There is a steady growth in the performance of mean percent (%) improvement for both Gaussian and KLD diversity measures as shown in Figure 9.5. It must be noted that we have considered only the degraded and improved performances and not the cases where the newly obtained accuracies are similar to the original ones (i.e., non-changed cases with 0% improvements). The graphical representation of performance for degraded and non-degraded cases is displayed in Figure 9.6. The best #Mo5 improvements obtained from all the four diversity measures irrespective of the m_{use} size are summarized in Figure 9.7. The highest improvement obtained is as high as 7.14% where the original accuracy (acc1=78.57%) is improved to a new accuracy (acc2=85.71%). A comparative study of a few best performing classification techniques on the dataset DS-IV is given in Table 9.14.

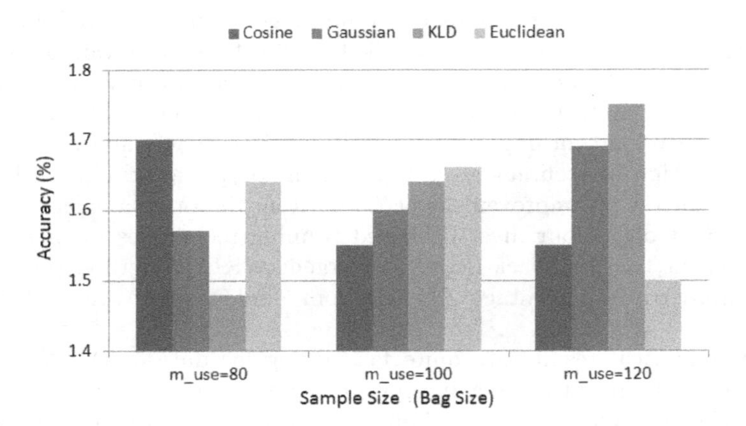

FIGURE 9.5: Mean percent (%) improvement obtained using different bag sizes from dataset DS-IV.

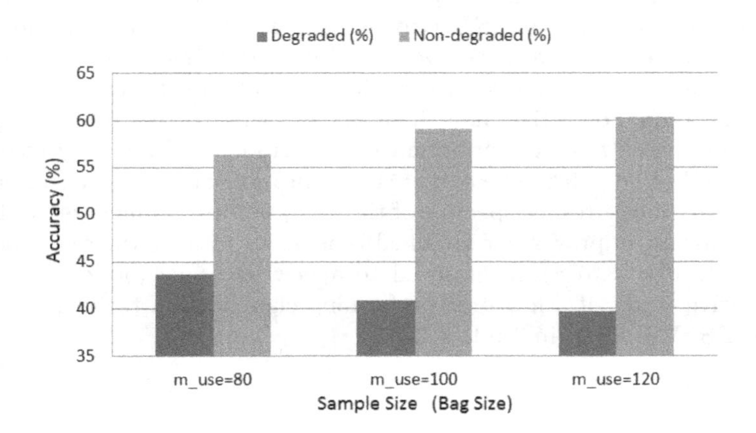

FIGURE 9.6: Mean degraded and non-degraded cases obtained using different bag sizes from dataset DS-IV.

TABLE 9.9: Configuration of the mix-bagging ensemble classifier.

Predictors#	Base Classifiers	Parameters
P1	Tree	Ovabyclass
P2	KNN	K=13
P3	Discriminant	Linear
P4	Naive Bayes	-
P5	SVM	Linear

TABLE 9.10: Results obtained from 5 independent runs of 100 times with $m_{use} = 45$ from DS-III.

Diversity Measures	Degraded (%)	Improved (%)	Mean (%) Improved	Mean #Mo5 (%)	#Mo5 (%)
Cosine	16.6	23.0	2.70	4.50	5.00
Gaussian	18.6	23.0	2.65	4.50	5.00
KLD	4.8	42.4	2.84	6.00	7.50
Euclidean	14.8	23.6	2.76	5.00	7.50
Mean±Std.	13.70±6.13	28.0±9.60	2.74±0.08	5.00±0.71	6.25±1.44

TABLE 9.11: Results obtained from 5 independent runs of 100 times with $m_{use} = 80$ from DS-IV.

Diversity Measures	Degraded (%)	Improved (%)	Mean (%) Improved	Mean #Mo5 (%)	#Mo5 (%)
Cosine	44.2	39.6	1.70	4.57	5.71
Gaussian	40.2	46.0	1.57	4.58	5.71
KLD	45.8	38.2	1.48	4.43	5.00
Euclidean	44.4	41.0	1.64	4.28	5.71
Mean±Std.	43.65±2.41	41.20±3.40	1.60±0.10	4.47±0.14	5.53±0.36

9.6 Conclusion and Future Work

The proposed method for performance improvement is a simple approach which does not involve further training or exhaustive combining techniques to enhance its performance. It uses the subset of already obtained prediction results and provides a substantially improved performance. As we discussed in the last section, irrespective of the different diversity measures, the obtained results are quite stable and similar for motor-imagery EEG dataset DS-IV (1.60% to 1.62% and 4.47% to 4.89% mean percentage improvement and mean #Mo5, respectively). The use of high bag size increases the probability for

TABLE 9.12: Results obtained from 5 independent runs of 100 times with $m_{use} = 100$ from DS-IV.

Diversity Measures	**Degraded (%)**	**Improved (%)**	**Mean (%) Improved**	**Mean #Mo5 (%)**	**#Mo5 (%)**
Cosine	40.8	44.4	1.55	4.86	6.43
Gaussian	39.2	45.6	1.60	4.57	5.71
KLD	39.8	45.0	1.64	5.28	7.14
Euclidean	43.8	41.2	1.66	4.86	6.43
Mean±Std.	40.90±2.04	44.05±1.96	1.61±0.05	4.89±0.29	6.28±0.60

TABLE 9.13: Results obtained from 5 independent runs of 100 times with $m_{use} = 120$ from DS-IV.

Diversity Measures	**Degraded (%)**	**Improved (%)**	**Mean (%) Improved**	**Mean #Mo5 (%)**	**#Mo5 (%)**
Cosine	40.4	44.8	1.55	4.43	5.71
Gaussian	38.8	45.4	1.69	4.58	5.00
KLD	38.5	45.6	1.75	4.57	5.00
Euclidean	41.0	46.8	1.50	4.86	6.43
Mean±Std.	39.68±1.21	45.65±0.84	1.62±0.12	4.61±0.18	5.54±0.68

TABLE 9.14: Comparison of a few best performing classification techniques on the dataset DS-IV.

References	**Used Features**	**Classifiers**	**Accuracy (%)**
[24]	Common Spatial Pattern	Adaboost	73.42
[11]	Band Power	Adaboost	83.57
[2]	Average Band Power	KNN	84.29
[7]	Energy-entropy	SVM	85.00
Proposed Algorithm 2 (KLD based diversity)	Energy-entropy	Mix-bagging	**85.71**

getting improvement in accuracy. However, if we consider the non-degraded (combined percentage of improved and non-changed) cases, it is always higher than that of the degraded cases independent of the experimental configuration. We can conclude that the proposed diversity matrix based pruning technique for performance improvement in ensemble learning performs quite satisfactorily in its first adaptation.

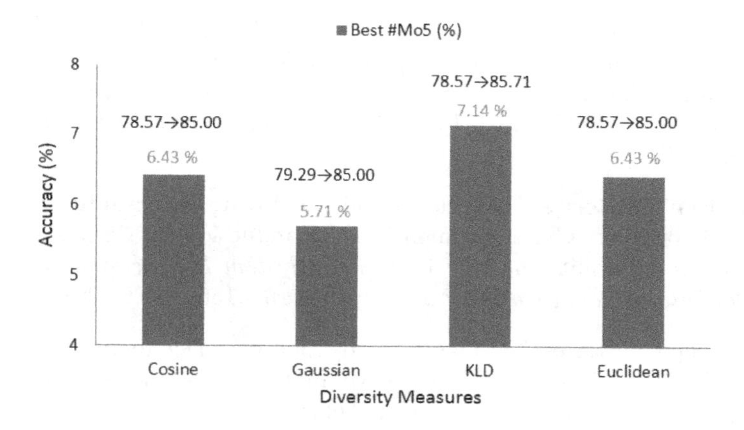

FIGURE 9.7: Best #Mo5 improvements ($acc1 \rightarrow acc2$) obtained from four diversity measures using the dataset DS-IV.

In the future, new strategies can be developed for an adaptive bag size so that it can maximize the improvement percentage and also restrict the proposed method from obtaining negative performance. Also, the study can be extended to develop a generic method which could work on different multi-class (more than two classes) datasets as well.

Bibliography

[1] BCI-Competition-II. *Dataset III, Department of Medical Informatics, Institute for Biomedical Engineering, University of Technology Graz*, Jan. 2004 (accessed June 6, 2015).

[2] Saugat Bhattacharyya, Anwesha Khasnobish, Somsirsa Chatterjee, Amit Konar, and DN Tibarewala. Performance analysis of lda, qda and knn algorithms in left-right limb movement classification from EEG data. In *Systems in Medicine and Biology (ICSMB), 2010 International Conference on*, pages 126–131. IEEE, 2010.

[3] Leo Breiman. Bagging predictors. *Machine learning*, 24(2):123–140, 1996.

[4] Rajdeep Chatterjee and Tathagata Bandyopadhyay. EEG based motor imagery classification using SVM and MLP. In *Computational Intelligence and Networks (CINE), 2016 2nd International Conference on*, pages 84–89. IEEE, 2016.

[5] Rajdeep Chatterjee, Tathagata Bandyopadhyay, and Debarshi Kumar Sanyal. Effects of wavelets on quality of features in motor-imagery EEG signal classification. In *Wireless Communications, Signal Processing and Networking (WiSPNET), International Conference on*, pages 1346–1350. IEEE, 2016.

[6] Rajdeep Chatterjee, Tathagata Bandyopadhyay, Debarshi Kumar Sanyal, and Dibyajyoti Guha. Dimensionality reduction of EEG signal using fuzzy discernibility matrix. In *Human System Interactions (HSI), 2017 10th International Conference on*, pages 131–136. IEEE, 2017.

[7] Rajdeep Chatterjee, Tathagata Bandyopadhyay, Debarshi Kumar Sanyal, and Dibyajyoti Guha. Comparative analysis of feature extraction techniques in motor imagery EEG signal classification. In *Proceedings of First International Conference on Smart System, Innovations and Computing*, pages 73–83. Springer, 2018.

[8] Rajdeep Chatterjee, Ankita Datta, and Debarshi Kumar Sanyal. Ensemble learning approach to motor imagery EEG signal classification. *Machine Learning in Bio-Signal Analysis and Diagnostic Imaging*, pages 183–208, 2018.

[9] Rajdeep Chatterjee, Dibyajyoti Guha, Debarshi Kumar Sanyal, and Sachi Nandan Mohanty. Discernibility matrix based dimensionality reduction for eeg signal. In *Region 10 Conference (TENCON), 2016 IEEE*, pages 2703–2706. IEEE, 2016.

[10] Ankita Datta and Rajdeep Chatterjee. Comparative study of different ensemble compositions in EEG signal classification problem. In *Emerging Technologies in Data Mining and Information Security*, pages 145–154, Singapore, 2019. Springer Singapore.

[11] Ankita Datta, Rajdeep Chatterjee, Debarshi Kumar Sanyal, and Dibyajyoti Guha. An ensemble classification approach to motor-imagery brain state discrimination problem. In *Infocom Technologies and Unmanned Systems (Trends and Future Directions)(ICTUS), 2017 International Conference on*, pages 322–326. IEEE, 2017.

[12] Minh N Do and Martin Vetterli. Wavelet-based texture retrieval using generalized Gaussian density and Kullback-Leibler distance. *IEEE Transactions on Image Processing*, 11(2):146–158, 2002.

[13] RA Fisher and Michael Marshall. Iris data set. *RA Fisher, UC Irvine Machine Learning Repository*, 440, 1936.

[14] Tapan Gandhi, Bijay Ketan Panigrahi, and Sneh Anand. A comparative study of wavelet families for EEG signal classification. *Neurocomputing*, 74(17):3051–3057, 2011.

[15] Jiawei Han, Jian Pei, and Micheline Kamber. *Data mining: Concepts and Techniques*. Elsevier, 2011.

[16] Jianfeng Hu, Dan Xiao, and Zhendong Mu. Application of energy entropy in motor imagery EEG classification. *JDCTA*, 3(2):83–90, 2009.

[17] Anna Huang. Similarity measures for text document clustering. In *Proceedings of the Sixth New Zealand Computer Science Research Student Conference (NZCSRSC2008), Christchurch, New Zealand*, pages 49–56, 2008.

[18] Pari Jahankhani, Kenneth Revett, and Vassilis Kodogiannis. Data mining an EEG dataset with an emphasis on dimensionality reduction. In *Computational Intelligence and Data Mining, 2007. CIDM 2007. IEEE Symposium on*, pages 405–412. IEEE, 2007.

[19] Richard Jensen and Qiang Shen. Fuzzy-rough sets for descriptive dimensionality reduction. In *Fuzzy Systems, 2002. FUZZ-IEEE'02. Proceedings of the 2002 IEEE International Conference on*, volume 1, pages 29–34. IEEE, 2002.

[20] Sotiris B Kotsiantis, Ioannis D Zaharakis, and Panayiotis E Pintelas. Machine learning: a review of classification and combining techniques. *Artificial Intelligence Review*, 26(3):159–190, 2006.

[21] Nojun Kwak and Chong-Ho Choi. Input feature selection by mutual information based on parzen window. *IEEE Transactions on Pattern Analysis and Machine Intelligence*, 24(12):1667–1671, 2002.

[22] Changki Lee and Gary Geunbae Lee. Information gain and divergence-based feature selection for machine learning-based text categorization. *Information Processing & Management*, 42(1):155–165, 2006.

[23] David G Lowe. Similarity metric learning for a variable-kernel classifier. *Neural Computation*, 7(1):72–85, 1995.

[24] Mostafa Mohammadpour, MohammadKazem Ghorbanian, and Saeed Mozaffari. Comparison of EEG signal features and ensemble learning methods for motor imagery classification. In *Information and Knowledge Technology (IKT), 2016 Eighth International Conference on*, pages 288–292. IEEE, 2016.

[25] Diego SC Nascimento, Anne MP Canuto, Ligia MM Silva, and Andre LV Coelho. Combining different ways to generate diversity in bagging models: An evolutionary approach. In *Neural Networks (IJCNN), The 2011 International Joint Conference on*, pages 2235–2242. IEEE, 2011.

[26] Zdzisław Pawlak. Rough sets. *International Journal of Computer & Information Sciences*, 11(5):341–356, 1982.

[27] Zdzislaw Pawlak. Rough set theory and its applications to data analysis. *Cybernetics & Systems*, 29(7):661–688, 1998.

[28] Gert Pfurtscheller, Christa Neuper, Alois Schlogl, and Klaus Lugger. Separability of EEG signals recorded during right and left motor imagery using adaptive autoregressive parameters. *IEEE Transactions on Rehabilitation Engineering*, 6(3):316–325, 1998.

[29] J Ross Quinlan et al. Bagging, boosting, and c4. 5. In *AAAI/IAAI, Vol. 1*, pages 725–730, 1996.

[30] Akhlaqur Rahman and Sumaira Tasnim. Ensemble classifiers and their applications: A review. *arXiv preprint arXiv:1404.4088*, 2014.

[31] Lior Rokach. Ensemble-based classifiers. *Artificial Intelligence Review*, 33(1):1–39, 2010.

[32] Fabrice Rossi, Amaury Lendasse, Damien François, Vincent Wertz, and Michel Verleysen. Mutual information for the selection of relevant variables in spectrometric nonlinear modelling. *Chemometrics and Intelligent Laboratory Systems*, 80(2):215–226, 2006.

[33] Jonathon Shlens. A tutorial on principal component analysis. *arXiv preprint arXiv:1404.1100*, 2014.

[34] Katarzyna Stapor. Evaluating and comparing classifiers: Review, some recommendations and limitations. In *International Conference on Computer Recognition Systems*, pages 12–21. Springer, 2017.

[35] Jonathan R Wolpaw, Niels Birbaumer, Dennis J McFarland, Gert Pfurtscheller, and Theresa M Vaughan. Brain–computer interfaces for communication and control. *Clinical Neurophysiology*, 113(6):767–791, 2002.

[36] Lean Yu, Shouyang Wang, and Kin Keung Lai. Investigation of diversity strategies in SVM ensemble learning. In *Natural Computation, 2008. ICNC'08. Fourth International Conference on*, volume 7, pages 39–42. IEEE, 2008.

Index

Absolute mean, 133, 144
Accuracy, 211
active contours, 180, 182, 185, 190
AD-LBP, 136
adaptive Cartesian harmony search, 50
age estimation, 180, 181, 183, 185, 188
Algorithm, 202, 206
ALOT, 146
AMEM, 133, 142, 144
ant colony optimization, 35
Artificial neural network, 45, 114
Automatic TB detection, 66

Bagging, 201
balanced Cartesian genetic programming, 44
Bayesian methods, 68
BCI, 198
Big Data, 90
binary NI, 7
biogeography-based optimization, 32
Brodatz, 143
BSE Sensex, 155, 159, 173
butterfly, 9, 10

Cartesian ant programming, 43
Cartesian genetic programming, 37
cgp-aco, 43
cgp-pso, 51
cgpann, 46
cgpde, 46
chaotic NI, 7
chemotherapy, 9, 15, 16
Classification, 200
CLBP, 138
CLBP_M, 138
CLBP_S, 138
Collaborative Filtering, 85
Compressive strength, 113
Conventional microscope, 64
Convolutional neural network, 71
correlation, 154, 155, 157, 163, 167, 169
Cosine Dissimilarity, 204

data vectors, 155
Database, 150
Dataset, 207, 210
Decision tree and Random forest, 69
detrending, 158
differential evolution, 32
Dimension reduction, 199, 200
Discernibility matrix, 199
Diversity matrix, 205
DWT, 200

Edge, 133, 134, 140, 144
EEG, 198
EMH, 154
Energy measurement filter, 133, 140, 142, 145
Energy-entropy, 201
Ensemble, 198
Ensemble learning, 201
Entropy, 133, 142, 144
estimation of distribution algorithms, 35
Euclidean Distance, 204
evolution strategies, 31
evolutionary algorithms, 31
Experiments, 209, 210

Feature selection, 199

feature space, 167, 171
floating-cgp-eda, 48
floating-point cgp, 47
Fluorescent microscope, 64
Fourier Transform, 199
Fuzzy classification, 70

Gaussian Dissimilarity, 204
genetic algorithms, 33
genetic programming, 37
GIS, 90
grasshopper, 9–12

harmony search, 34
hybrid Cartesian genetic programming algorithms, 42
hybrid evolutionary strategies, 49
hybrid metaheuristics, 40
Hybrid methods, 71
hybrid NI, 9

Intensity LBP, 136

k-NN, 133
KPCA, 155, 159, 166, 167, 169, 171, 175
Kullback-Leibler Divergence, 204

Laws' mask, 133
LBP, 133
LBP intensity, 136
LBP median, 132, 137
LBP minimum, 136, 150
LBP number, 137, 150
LBP uniform, 136, 150
LBP variance, 132, 140
Level, 142
LPLEM, 140–142

Machine Learning, 92
Machine learning techniques, 68
Majority voting, 202
MEM, 142, 144, 145
metaheuristics, 30
Metallic biomaterials, 102
Micro-hardness, 111
min-max, 133, 140
Mix-bagging, 198, 202
Mo5, 210
Modified Random Perturbation, 95
Motor-imagery, 199
Moving window, 142
multi-objective NI, 8
Multiobjective, 33

Naive Baysian Classifion, 94
nature-inspired, 2, 4
Neural Networks, 70
neuroevolution, 46
NNM Factorization, 94
No-Free-Lunch theory, 2
noise, 162, 169, 175
non-dominated sorting genetic algorithm, 33
Normalization, 133, 140, 142, 144

opposition-based learning, 45
orthopaedic application, 103

particle swarm optimization, 36
PCA, 155, 159–163, 175, 199
Pre-processing, 199
Privacy, 94
Pruning, 205

RD-LBP, 136, 150
Results, 210, 211
Ripple, 133, 142
Rough-set, 199

S/N ratio, 113
salp, 9, 12, 13
SDEM, 133, 142, 144, 145
sheep age, 180, 188, 191, 194
Singular Value Decomposition, 97
Sobel_LBP, 132, 139, 148
Spot, 140, 144
spotted hyena, 9, 14, 15
Support Vector Machine, 70
SVM, 198
swarm intelligence, 2–5

Taguchi method, 109
TB bacilli, 64
teeth segmentation, 187, 194
Texture energy measures, 133, 142
Texture parameters, 133, 140, 142
Touching bacilli, 65
Tourism Recommender System, 82
Tuberculosis, 63

Uniformity measure, 136
univariate marginal distribution algorithm, 48

Vectors, 140

Wavelet Transform, 200
Wear measurement, 107
Window size, 142

Ziehl-Neelsen, 64